Newnes
Service
Engineer's
Pocket Book

Geoff Lewis and Ian Sinclair

Newnes
An imprint of Butterworth-Heinemann
Linacre House, Jordan Hill, Oxford OX2 8DP
225 Wildwood Avenue, Woburn, MA 01801–2041
A division of Reed Educational and Professional Publishing Ltd

 A member of the Reed Elsevier plc group

OXFORD   BOSTON   JOHANNESBURG
MELBOURNE   NEW DELHI   SINGAPORE

First published 1998

British Library Cataloguing in Publication Data
A catalogue record for this book is available from the British Library

ISBN 0 7506 3448 0

Library of Congress Cataloguing in Publication Data
A catalogue record for this book is available from the Library of Congress

Printed in Great Britain

CONTENTS

FOREWORD

The problems associated with servicing domestic electronic equipment, and hence the need for reliable guidance, has never been greater. Today's service technician has to deal with a vast range of equipment that comes from all parts of the globe. He has, therefore, to be fully versed in the relevant electronic circuitry and able to think in terms of systems – what effect might produce a particular symptom. Digital control has added an extra dimension to this. An appreciation of such things as the effects of heat, the need for adequate ventilation, the effects of mechanical stress, the quality of components and their specification, the quality of assembly and in particular soldering is essential in being able to assess fault situations and suss out likely causes.

In addition to the technical problems there are the commercial ones. Servicing today has to be carried out in a tight financial context. For servicing to be economically viable, it is necessary to be able to carry out fault diagnosis and complete the necessary repair work speedily. Reliable repairs are also essential: bounces and call-backs can have a devastating effect on the viability of a servicing organization.

Most textbooks for the electronics technician concentrate on components and circuitry. This book takes the technician a stage further, to help him (or her) deal with the sorts of problems that arise in practice when faulty equipment has to be repaired. It thus fulfils an important role and should, in particular, help those aiming for an NVQ or GNVQ. It will also serve as a useful reference source. The authors have given careful thought to the presentation of the information in this book, and have unrivalled experience in writing for the electronics technician.

The book is a welcome and important addition to the literature available to today's service technicians.

John A. Reddihough
Editor, *Television*
1997.

PREFACE

When presented with the proposal to write an all-embracing book aimed at the servicing of domestic entertainment, the authors were faced with the task of trying to squeeze many aspects into a single volume. It was therefore decided to introduce each topic through short revisionary notes to explain how the system works, so that fault conditions and repair methods could be discussed at the relevant points in the text. It has been assumed that the reader would already be qualified to the Part 2 level of C&G 2240 or had acquired a similar level of practical competence and was also conversant with the material included in such books as *Servicing Electronic Systems* published by Ashgate Publishing Co.

Historically, most service departments were supported by the profits from a strong sales department. However, in today's trading, each branch is expected to be self-supporting. It is therefore the major aim of the service department to create satisfied customers who will not only return, but will spread your reputation by personal recommendations. Such self creating advertising is free of charge. To achieve this desirable state, it is important that the workshop should be adequately equipped and well organized. This involves not only keeping a good library of service manuals and data sheets, but having access to as many other aids as possible. The authors feel that perhaps one of the best aids is a subscription to a journal such as *Television*, which acts very much as a clearing house for the hints, tips and solutions to problems encountered by the many practising service engineers.

It is a fact of life that today's service engineer is often presented with a problem after a knowledgeable owner has attempted a first line attack. One particularly good example of this is the use of cassette cleaner tapes for both audio and video machines. The owner is not often very pleased to find that after purchasing one of these, it is defunct after about half a dozen passes and starts to create more troubles than existed in the first place. Again, quite often a cassette jams in a VCR and the owner tries to prise a valuable recording (to him) out of the machine with a screwdriver. All more grist to servicing mill, but it does not help the customer's temper to be told he should not have approached the problem this way. It was recently reported that a certain freezer spray that is used in the workshop when tackling temperature-sensitive faults can generate a high level of static electricity! A further task for the servicing staff is to try to find time to read as many of the technical papers that are continually being produced.

The physical handling of the new digital television receivers with wide screen 16:9 pictures and 42 inch CRTs, will need considerable care to avoid falling foul of the HSE regulations. Maybe the early development of the solid state hang-on-the-wall, display devices will reduce the problems of back-strain but at the same time provide more work for the service engineers. The future of servicing has always been obscured by technical developments, but how much more work is going to be generated by the Internet communicating television set?

The authors wish to recognize the global support that has been made available to us by the test equipment, set and chip manufacturers. Without this help the project would not have been possible.

Geoff Lewis and Ian Sinclair, 1997

CHAPTER 1

GENERAL FAULTFINDING

Faultfinding for electronics equipment is a skill that is neither an art nor a science, but an engineering discipline in its own right. Effective faultfinding requires:

- A good general knowledge of electronics.
- Specialized knowledge of the faulty equipment.
- Suitable test equipment.
- Experience in using such test equipment.
- The ability to formulate a procedure for isolating a fault.
- The availability of service sheets and other guides.

A good general knowledge of electronics is essential because not all equipment is well documented, and in some cases only a circuit diagram (or even nothing at all) may be available as a guide. Failing a concise description of how the equipment works, you may have to work out for yourself the progress of a signal through the equipment. In addition, wide general knowledge is needed if you are to make reasonable assumptions about how to substitute components. At the risk of being obvious, you are not likely to know why something doesn't work if you don't know what *does* make it work.

Specialized knowledge can greatly reduce the time spent in servicing, and if your servicing is confined to a few models of equipment you are likely to know common or recurring faults by their symptoms. All too often, however, service engineers are likely to have to struggle with unfamiliar equipment for a large portion of their time.

Suitable test equipment is essential. The days when a service engineer could function effectively with little more than an Avometer and a screwdriver are long gone, and though the multimeter is still an important tool (as also is the screwdriver) the service engineer needs at least one good general-purpose oscilloscope, along with signal generators, pulse generators, and more specialized equipment appropriate for the type of equipment he/she is working on.

Experience in using test equipment is also an essential. All test instruments have limitations, and you must know what these are and how you can avoid being hung up by these limitations. You must know which tests are appropriate for the faulty equipment, and what the result of such tests would be on equipment that was not faulty.

The ability to formulate a procedure for isolating a fault means that you need to know what to test. All electronic equipment consists of sections, and much modern equipment uses a single IC per section. You should be able to pin down a fault to one section in a logical way, so that you do not waste time in performing tests on parts of the circuit which could not possibly cause the fault. The classical method of isolating a fault has, in the past, been to check signal inputs and outputs for each stage, but this is no longer the only method that needs to be used, and in some cases, the use of feedback loops, limiters, and other interacting circuits makes it much more difficult to find where a fault lies. Once again, experience is a valuable guide

The availability of service sheets and other guides is also important. Much commercial equipment consists of components which carry only factory codes, and whose actions you can only guess at in the absence of detailed information. In addition, good service sheets will often carry a list of known recurring faults, and will also give valuable hints on faultfinding methods.

Finding a fault is not, unfortunately, a certain step towards repair. Some equipment carries ICs which are no longer in production and for which no replacement is available. Many firms, particularly manufacturers of domestic electronic equipment, will provide spares and help for only a limited period, and some firms seem to deny all responsibility for what their equipment does after a few years. Given the comparatively long life of most electronic equipment, it would be unreasonable to expect spares to be available indefinitely, but it is not easy to tell a customer that the TV receiver he/she bought only six years earlier cannot now be serviced because it contains parts for which there is no current equivalent. Manufacturers might like to remember that customers tend to have long memories about such things – it certainly affects my judgement when I want to buy anything new.

HEALTH AND SAFETY CONSIDERATIONS

The Health and Safety at Work Act of 1974 came into full operation in the UK in October 1978, and it affects all aspects of servicing, as it affects all other workshop activities. The Act is a logical expansion of the various Factory Acts up to and including 1961, and the accident logging procedures that are required are detailed in the Notification of Accidents and General Occurrences Regulations of 1980.

The core of the legislation is that any work must be carried out in a way that ensures maximum safety for the employee and for anyone else present. Employers must provide a safe working environment and see that safety rules are obeyed. All employees (including students or apprentices) have a duty to observe the safety precautions that have been laid down by their employers, and to carry out all work in a safe way.

The intentions of the Act can be summarized:

1. All employers, including teachers, tutors and instructors, must ensure that any equipment to be used or serviced by employees, apprentices, or students is as safe to operate as it can reasonably be made. Equipment which cannot be made safe should be used only under close supervision.
2. Employers, tutors and instructors must make sure that employees and students at all times make use of the required safety equipment such as protective clothing, goggles, ear protectors, etc. which the employer must provide for them.
3. Employees and students must ensure that they carry out all their work in such a way that it does not endanger them, people working around them, or the subsequent users of equipment that they repair.
4. Any accidents must be logged and reported, and measures taken to ensure that such incidents cannot happen again.
5. Failure to observe the provisions of the Act is reasonable grounds for dismissal.

Accident prevention depends on:

- recognition of hazards,
- elimination, if possible, of hazards,

- replacement of hazards,
- guarding and/or marking hazards,
- a sense of personal protection,
- continuing education in safety.

One particular difficulty that faces anyone working on servicing domestic electronics equipment is the wide variety of working places, which can vary from a well-designed workshop, as safe as possible in the light of current experience, to a home in which a TV receiver has failed, and in which almost every possible hazard, from bad electrical wiring to the presence of children around the TV receiver, exists. This huge difference in working conditions makes it vital for anyone working on electronics servicing to be aware of safe working methods and to practise them at all times whether supervised or not.

The hazards which lead to accidents can, as far as electronic servicing is concerned, be classed as:

- Electrical, particularly where industrial equipment is serviced.
- Fire, particularly where flammable materials are used.
- Asphyxia, particularly because of degreasing solvents.
- Mechanical, particularly in cramped conditions.

All of the current health and safety legislation emphasizes accident prevention as well as dealing with accidents. Dealing with accidents is a matter of provision of facilities for, mainly, first aid and fire fighting and is very much in the hands of the employer. Prevention of accidents depends on attitudes of mind and is the responsibility of everyone concerned.

These attitudes can, and must, be cultivated. The person who recognizes a hazard is the person who can imagine what might happen if an emergency arose, or who knows how to look for danger, and the Act implies that this person must not be treated as a troublemaker or whistle-blower, but as a valuable contributor to safety who can save money in the long run.

AREAS OF RESPONSIBILITY

The safety legislation requires both employer and employee to observe, maintain and improve safety standards and the adoption of safe working practices. Self-employed persons are responsible for their own safety and also the safety of anyone who works alongside them, and they also have a responsibility to avoid endangering members of the public. The self-employed should ensure that they have insurance protection against third party claims from the public, and for work outside the UK they must ensure also that their qualifications entitle them to be carrying out the grade of work which they are undertaking.

The responsibilities of an employer are:

1. To ensure that the workplace is structurally secure. Typical hazards are insecure floors, leaking roofs, blocked windows, restricted doorways and so on.
2. To provide safe plant and equipment. Work benches and equipment must be adequate for the job. All tools must be fitted with safety guards. Larger power tools should be screened or fenced. If necessary, floors should be marked with safe walking areas.
3. To lay down safe working systems. Employees do not have to use makeshift equipment or methods. Protective clothing should be provided if needed.

4. To ensure that environmental controls are used. The workshop must be kept at a reasonable working temperature. Humidity levels should be controlled if needed, and ventilation must be adequate. Employees should not be subjected to dust and fumes, and there must be washing facilities, sanitation for both men and women, and provision for first aid in the event of an accident.
5. To ensure safety in the handling, storing and movement of goods (see below). Employees should not be required to lift heavy loads (a useful guide, enforced by law in some EEC countries, is that an employee cannot be expected to lift without assistance a load of more than 15 kg). Mechanical handling must be provided where heavy loads are commonplace. Dangerous materials must be identified and stored where they do not cause any hazard to anyone on the site.
6. To provide a system for logging and reporting accidents. Such a log is not to be simply a list of happenings, but a guide to better work practices.
7. To provide information, training and updating of training in safety precautions, along with supervision that will ensure safety.
8. To devise and administer a safety policy that can be reviewed by representatives of both employer and employees.

MANUAL HANDLING

Many items of electronic equipment are heavy, some are awkwardly shaped, and some are deemed too fragile for mechanical handling unless suitably packaged. The following points are important when goods have to be shifted by hand.

- Plan the move, looking for any aids (a webbing strap, a strong blanket) that might help. Summon help if necessary.
- Remove obstructions and consider if you need to rest the load at an intermediate point such as a bench so that you can change your grip.
- Place your feet apart and put one foot as far forward as you can.
- Bend your knees (but do not kneel), and keep your back straight. Try to keep your shoulders level and in line with your hips because twisting is a certain way to strain the back.
- Try to keep the load as close to you as possible.
- Lift smoothly and do not swivel at your waist – if you need to turn, use your feet.
- If you need to place the load in some precise position, put it down first and adjust it later, by sliding it if possible.

If manual handling is a large part of the workshop duties, all employees and the employer should read the Health and Safety Executive booklet titled *Manual Handling: guidance on Regulations,* ISBN 0 11 886335 5 (HMSO 1992). This can be obtained from:

HSE Books,
PO Box 1999,
Sudbury,
Suffolk CO10 6FS
Tel: (01787) 881165
Fax: (01787) 313995

OUTSIDE WORK

When servicing has to be carried out in domestic premises or any other situation away from the workshop, the following guidelines are useful:

(a) ensure that servicing is carried out by staff who are adequately qualified and experienced (this is a legal necessity in Germany and Holland);
(b) ensure that more than one person is present if this would contribute to added safety;
(c) make full use of replaceable subassemblies so that detailed work can be carried out in the workshop rather than at the site.

The hazards of servicing in the home include:

- Unsound wiring, with poor earthing, and possibly overloaded cables.
- No fire precautions.
- Badly arranged working space.
- The presence of children.
- Bulky and heavy domestic equipment which has to be moved.

Unless a servicing job on domestic equipment is particularly simple and can be carried out quickly with the minimum of dismantling, it is always better to remove the equipment (even if this presents a lifting hazard) to work on at the workshop.

EMPLOYEES' RESPONSIBILITIES

The law recognizes that an employer can only provide a framework for safe working, and that it can be difficult to force an employee to work in a safe way. The Act therefore emphasizes that the employee also has responsibilities to ensure safe working. This is not just a matter of personal safety, but the safety of fellow workers and members of the public. Most accidents are caused by human carelessness in one form or another and though standards can be drawn up for safe methods of working, it is impossible to ensure that everyone will abide by these standards at all times.

The employees' responsibilities, which apply also to the self-employed, include:

Health care.
This means, for example, avoiding the use of alcohol or drugs if their effects would still be present in working hours. Remember that the term 'drugs' can include such items as painkillers, antihistamines for hay fever, and antidepressants. In some countries, blood alcohol levels can be checked following industrial accidents as well as following a road accident. Working while excessively tired can also be a cause of accidents due to relaxed vigilance.

Personal tidiness.
Clothing should provide reasonable protection, with no loose materials that can be caught in machinery or even (as has been reported) melted and set alight by a soldering iron. Long hair is even more dangerous in this respect, and must be fastened so that it cannot be caught.

Behaviour.
Carelessness and recklessness cannot be tolerated, and practical jokes, no matter how traditional, do not belong in any modern workplace. Legislation in several countries now treats this type of behaviour in the workplace as seriously as it is treated on the roads.

Competence.
An employee must know, either from experience, discussion with a colleague, or from reference to manuals, what has to be done and the safe way of carrying out the work. In several EEC countries, it is an offence to carry out work for which you are not qualified, and if you are working abroad you will have to find out to what extent UK qualifications apply in other countries.

Deliberately avoiding hazards.
Self-employed service personnel should insure against third party claims, and employees must remember that they can be sued if their careless or reckless behaviour leads to injury. Misusing safety equipment is a criminal offence, quite apart from endangering the lives of others. It is also an offence to fail to report a hazard which you have discovered.

ELECTRICAL SAFETY

The main electrical hazard in servicing operations is working on live equipment, because of the ever-present risk of shock. Electric shock is caused by current flowing through the body, and it is the amount of current and where it flows that is important. The resistance of the body is not constant so that the amount of current that can flow for a given voltage will vary according to the moistness of the skin. The most important hazard is of electric current flowing through the heart. Ways of avoiding this include the following:

1. Ensure that only low voltages (less than 50 V) are present in a circuit.
2. Ensure that no currents exceeding 1 mA can flow in any circumstances.
3. Keep the hands dry at all times, because moist hands conduct much better than dry hands.
4. Keep workshop floors dry, and wear rubber soled shoes or boots. Avoid two-handed actions, particularly if one hand can touch a circuit and the other hand is touching a metal chassis, metal bench or any other metal object.

The greatest hazard in most electronic servicing is the mains supply which in the UK and in most of Europe is at 240 V a.c. The correct use of three-pin plugs and sockets, correctly wired and fused, is essential. As a further precaution, earth leakage circuit breakers (ELCBs) can be used to ensure that any small current through the earth line will operate a relay that will open the live connection, cutting off the supply.

If live working is unavoidable, try to avoid the possibility of current passing through the heart in the following ways:

- Work with the right hand only, making a longer electrical path to the heart in the event of touching live connections.
- Wear insulating gloves.
- Cover any metalwork which is earthed.

Power tools present a hazard in most workshops, and must be electrically safe. Some useful points are:

- Always disconnect when changing speed or drill bits.
- Never use a power tool unless the correct guards and other protective devices have been correctly fitted.
- Ensure that the mains lead is in perfect condition, with no fraying or kinking, and renew if necessary.
- Metal tools should be earthed, and in some conditions can be powered from isolating transformers. Another option is to use only battery-powered tools.
- Most domestic power tools use plastic casings with double insulation, but these are not necessarily suitable for industrial uses.
- Test equipment must itself be electrically safe, using earthing or double insulation as appropriate.
- Test equipment can often be bought in battery-operated form.
- Mains-powered test equipment must use the correct mains voltage setting, have power leads in good condition and the fusing correct for the load (usually 3 A).
- All test equipment, particularly if used on servicing live equipment, should be subject to safety checks at regular intervals.
- Users should maintain such equipment carefully, avoiding mechanical or electrical damage.

Note: Low-power industrial equipment often makes use of standard domestic plugs, but where higher power electronic equipment is in use the plugs and socket will generally be of types designed for higher voltages and current, often for 3-phase 440 V a.c. In some countries, flat two-pin plugs and sockets are in use, with no earth provision except for cookers and washing machines, though by the end of the century uniform standards should prevail in Europe, certainly for new buildings.

Any mains electrical circuit should also include:

(a) A fuse or contact breaker whose rating matches the maximum consumption of the equipment. This fuse may have blowing characteristics that differ from those of the fuse in the plug. It may, for example, be a fast-blowing type which will blow when submitted to a brief overload, or it may be of the slow-blow type that will withstand a mild overload for a period of several minutes.
(b) A double-pole switch that breaks both live and neutral lines. The earth line must *never* be broken by a switch.
(c) A mains warning light or indicator which is connected between the live and neutral lines.

All these items should be checked as part of any servicing operation, on a routine basis. In addition:

- As far as possible all testing should be done on equipment that is disconnected and switched off.
- The absence of a pilot light or the fact that a switch is in the OFF position should never be relied on as a sign that it is safe to touch conductors.
- Mains-powered equipment being repaired should be completely isolated by unplugging from the mains.
- If the equipment is permanently wired then the fuses in the supply line must be removed before the covers are taken from the equipment.

Many pieces of industrial electronics equipment have safety switches built into the covers so that the mains supply is switched off at more than one point when the covers are removed.

The ideal to be aimed at is that only low-voltage battery-operated equipment should be operated on when live. When mains-powered equipment must unavoidably be tested live, the meter, oscilloscope or other instrument(s) should not be connected until the equipment is switched off and isolated, and the live terminals should be covered before the supply voltage is restored. The supply line should be isolated again before the meter or other instrument clips are removed.

The main dangers of working on live circuits are:

(a) The risk of fatal shock through touching high voltage exposed terminals which can pass large currents through the body.
(b) The risk of fatal shock from the discharge of capacitors that were previously charged to a high voltage.
(c) The risk of damage to instruments or to the operator when a mild electric shock is experienced. The uncontrollable muscular jerking which is caused by an electric shock of any kind can cause the operator to drop meters, to lose his/her balance and fall on to other, possibly more dangerous, equipment.

Even low-voltage circuits can be dangerous because of the high temperatures which can be momentarily generated when a short circuit occurs. These temperatures can cause burning, sometimes severe. As a precaution, chains, watchstraps and rings made of metal should not be worn while servicing is being carried out.

Older types of TV produced in the UK used 'live chassis' circuits in which the neutral lead of the mains supply was connected to the metal chassis of the equipment. Incorrect wiring of the plug or a disconnected neutral lead would cause the chassis to be live at full supply voltage. Later models of TV receiver used considerably lower internal voltages, but some still used a live chassis approach, and a few were wired so that the chassis was at about half of supply voltage, irrespective of how the plug was connected. Live working on such equipment should never be carried out without the use of an isolating transformer with the chassis securely earthed. Modern TV receivers will normally use a switch-mode power supply which is isolated from the mains so that SCART connectors can be used between the receiver and other equipment.

Other electrical safety points are:

- Cables should not be frayed, split, or be sharply bent, too tightly clamped, or cut.
- Damaged cables must be renewed at once.
- Hot soldering irons must be kept well away from cable insulation.
- Cables must be securely fastened both into plugs and into powered equipment. The supply cable to a heavy piece of equipment should be connected by way of a plug and socket which will part if the cable is pulled.
- The live end of the connector must have no pins which can be touched.
- Every electrical joint should be mechanically as well as electrically sound. It must be well secured with no danger of working loose and with no stray pieces of wire.
- The old-fashioned type of lead/acid accumulator releases an explosive mixture of hydrogen and oxygen while it is on charge, and this can be ignited by a spark. No attempt should ever be made to connect or disconnect such accumulators from a charger

or from a load while current is flowing. An explosion is serious enough, but the explosive spray of acid along with sharp glass or plastic fragments is even worse.

FIRE PRECAUTIONS

Workshop fires present several types of hazard. The obvious one is flesh burns, but injuries from other causes are common, such as falling when running from a burning workshop. Clear escape routes should be marked and these must be kept clear at all times. If there is further risk of suffocation by smoke, low-level emergency lighting should be present so that a route to a safe exit is marked.

A fire can start wherever there is material that can burn (combustible material), air, or oxygen-rich chemicals, that can supply oxygen for burning, and any hot object that can raise the temperature of materials to the burning point. A fire can be extinguished by removing all combustible material, by removing the supply of air (oxygen), by smothering with non-combustible gas or foam, or by cooling the material to below the burning temperature. The most dangerous fires are those in which the burning material can supply its own oxygen (such materials are classed as *explosives*) and those in which the burning material is a liquid that can flow about the workplace taking the fire with it.

Another major danger is of asphyxiation from the fumes produced by the fire. In electronics workshops the materials that are used as switch cleaners, the wax in capacitors, the plastics casings, the insulation of transformers and the selenium that can still be found in some old-fashioned metal rectifiers will all produce dangerous fumes, either choking or toxic, when burning. Good ventilation can reduce much of this particular hazard.

The key points for fire safety are:

(a) Good maintenance.
(b) No naked flames and no smoking permitted.
(c) Tidy working, with no accumulation of rubbish.
(d) Clearly marked escape paths in case of fire.
(e) Good ventilation to reduce the build-up of dangerous fumes.
(f) Suitable fire extinguishers in clearly marked positions.
(g) Knowledge by all staff of how to deal with fire/explosion.
(h) Regular fire drills and inspection of equipment.

Because of the variety of materials that can cause a fire, more than one method of extinguishing a fire may have to be used, and use of the wrong type of extinguisher can sometimes make a fire worse. The five main types of fire extinguisher and their colour coding are:

- water based (**red**),
- foam (**cream**)
- powder (**blue**),
- CO_2 (carbon dioxide) gas (**black**),
- inert liquid (**green**).

The water-based type of extinguisher (for a **Class A** fire) is most effective on materials that absorb water and which will be cooled easily. Fires in paper, wood or cloth are best tackled in this way, using the extinguisher on the base of the fire to wet materials that are not yet burning and to cool materials that have caught light. Water-based extinguishers must *never* be used on electrical fires or on fires that occur near to electrical equipment.

Class B fires involve burning liquids or materials that will melt to liquids when hot. The main hazards here are fierce flames as the heat vaporizes the liquid, and the ease with which the fire can spread so as to affect other materials. The most effective treatment is to remove the air supply by smothering the fire, and the foam, dry powder or CO_2 gas extinguishers can all be useful, though such extinguishers should be used so that the foam, powder or gas falls down on to the fire, because if you direct the extinguisher at the base of this type of fire the liquid will often simply float away, still burning. Fire blankets can be effective on small fires of this type, but on larger fires there is a risk that the blanket will simply act as a wick, encouraging the fire by allowing the burning liquid to spread.

For dealing with electrical fires, the water- or foam-based types should be avoided because the risk of electric shock caused by the conduction of the water or foam is often more serious than the effect of a small fire. Inert liquid extinguishers are effective on electrical fires but the liquids can generate toxic fumes and can also dissolve some insulating materials. Powder extinguishers and CO_2 types are very effective on small fires of the type that are likely to develop in electronics workshops. They must be inspected and checkweighed regularly to ensure that internal pressure is being maintained.

A few sand buckets are also desirable. They must be kept full of clean sand and never used as ashtrays or for waste materials. A firemat is also an important accessory in the event of setting fire to the clothing of anyone in the workshop. The firemat should be kept in a prominent place and everyone in the workshop should know how to use it.

In the event of a fire, your order of priorities should be:

1. To raise the alarm and call the fire brigade even if you think you can tackle the fire.
2. To try to ensure evacuation of the workshop.
3. To make use of the appropriate fire extinguishers.

The important point here is that you must never try to fight a fire alone, nor to put others at risk by failing to sound the alarm. The most frightening aspect of fire is the way that a small flame can in a few seconds become a massive conflagration, completely out of hand, and though prompt use of an extinguisher might stop the fire at an early stage, raising the alarm and calling for assistance is more important.

GENERAL SAFETY

Soldering

Soldering should never be carried out on any equipment, whether the chassis is live or not, until the chassis has been completely disconnected from the mains supply and all capacitors safely discharged. The metal tip of the soldering iron will normally be earthed, and should not be allowed to come into contact with any metalwork that is connected to the neutral line of the mains, because large currents can flow between neutral and earth. Obviously, any contact between the soldering iron and the live supply should also be avoided.

Soldering and desoldering present hazards which are peculiar to the electronics workshop. The soldering iron should always be kept in a covered holder to prevent accidental contact with hands or cables. Many electrical fires are started by a soldering iron falling on to its own cable or the cable of another power tool or instrument. Though it can be very convenient to hang an iron up on a piece of metal, the use of a proper stand with a substantial heat sink is the safe method that ought to be used.

In use, excessive solder should be wiped from the iron with a damp cloth rather than by being flicked off and scattered around the workshop. During desoldering, drops of solder should not be allowed to drip from a joint; they should be gathered up on the iron and then wiped off it, or sucked up by a desoldering gun which is equipped with a solder pump.

Care should be taken to avoid breathing in the fumes from hot flux. Where soldering is carried out on a routine basis, an extractor hood should be used to ensure that fumes are efficiently removed. Materials other than soldering flux can cause fumes, and some types of plastics, particularly the PTFE types of materials and the vinyls, can give off very toxic fumes if they are heated to high temperatures. Extraction can help here, but it is preferable to use careful methods of working which ensure that these materials are not heated.

Toxic materials

Virtually every workplace contains toxic materials, or materials that can become toxic in some circumstances, such as a fire. Industrial solvents, such as are used for cleaning electronics subassemblies, switch contacts, etc. are often capable of causing asphyxiation and even in small concentrations can cause drowsiness and stupor. Many common insulating materials are safe at normal temperatures, but can give off toxic fumes when hot.

In addition, some very poisonous materials are used within electronics components. Such materials can be found in cathode-ray tubes, fluorescent tubes, valves, metal rectifiers, power transistors, electrolytic capacitors and other items. The local hospital should be informed of the toxic materials that are present, and, if possible, should advise on any first aid that might be effective in the event of these materials being released. In particular, cathode-ray tubes and fluorescent tubes must be handled with care because they present the hazard of flying glass as well as of toxic materials if they are shattered.

Transistors and ICs should never in any circumstances be cut open. No servicing operation would ever call for this to be done, but if a faulty component has to be cut away there might be a danger of puncturing it. Some power transistors, particularly those used on transmitters, contain the solid material beryllium oxide whose dust is extremely poisonous if inhaled, even in very small quantities. All such transistors that have to be replaced should be removed very carefully from their boards and returned to the manufacturers for safe disposal.

Corrosive chemicals form another class of toxic material which cause severe damage to the skin on contact. Sulphuric acid has the effect of removing water from the skin, causing severe burning, whereas sodium hydroxide (caustic soda) dissolves fatty materials and thus damages the skin by removal of fat. All strongly acidic or alkaline materials will cause severe skin damage on contact, but by far the most dangerous in addition to sulphuric acid and sodium hydroxide are hydrofluoric acid (used for etching glass) and nitric acid (used in etching copper). The only first aid treatment is to apply large amounts of water to dilute the corrosive material. Even very dilute acids or alkalis will cause severe damage if they reach the eyes, and once again, large amounts of pure water should be applied as a first aid measure. *Never* attempt to neutralize one chemical with another, because neutralization is usually accompanied by the generation of heat. Specialized treatment at hospital should always be sought in any case of accident with corrosive materials.

Use of hand tools

The incorrect use of small hand tools is a very common cause of accidents and safe methods of working must always be used *especially* when you are in a hurry.

- Never point a screwdriver at your hand. This seems obvious, yet a common cause of stabbing accidents is using a screwdriver to unscrew a wire from a plug that is held in the palm of the hand rather than between fingers or, better still, in a vice.
- Small tools have to be suited to the job and used correctly. Never use a blade screwdriver when a Phillips or Pozidrive type is needed, and remember that screwdriver size must be matched to the work.
- Files should be fitted with handles and used with care because, being brittle, they can snap. Never use a file as a tommybar, for instance.
- Box spanners and socket sets should be used in preference to open spanners where possible.
- Snipping wire with sidecutters can be a hazard to the eyes, either of the user of the cutters or of anyone standing close. Ideally, the wire which is being snipped should be secured at both ends so that no loose bits can fly around. One safe method is to hold the main part of the wire in a vice and the other part in a Mole wrench.

Protective clothing should be worn wherever the regulations or plain common sense demands it. The workshop is no place for loose ties or cravats, for long, untied hair or strings of beads. A workshop coat or boiler suit should be worn whenever workshop tools are being used, and eyeshields or goggles are nearly always necessary. Goggles and gloves must always be worn when cathode-ray tubes are being handled and safety boots should be worn when heavy objects have to be moved. Goggles are also a useful protection when wire is being snipped, particularly for hard wire like nichrome.

Back injury

Back injury is the single most common cause of absence from work in the UK. In the course of electronics servicing work many large and heavy items have to be shifted. Injury is most commonly caused by incorrect methods of lifting heavy objects or holding such objects before putting them into place.

The correct method of lifting is to keep a straight back throughout the whole of the lifting operation, bending your knees as necessary but *never* your back.

- Never attempt to shift anything heavy by yourself.
- Even if you bear all the weight of carrying a load, help in lifting and steering the load can make the difference between safe and unsafe work.
- In several countries, there is a statutory upper limit of load that one operator is allowed to lift without help. This can be as low as 15 kg (33 lb), which is less than the usual 20 kg (44 lb) weight allowance for luggage on flights.

- Workshop benches and stools should be constructed so that excessive lifting is not required.
- If a large number of heavy items have to be moved to and from benches, mechanical handling equipment should be used.
- Stool heights should be adjustable so that no user needs to stoop for long periods to work on equipment.

Reporting hazards and accidents

An employer must ensure that a workspace is kept free of dangerous and badly sited materials. Someone in the workshop, however, may have put the hazard into place, and quite certainly someone who works near the hazard will be the first to notice it. The responsibility cannot be shrugged off as being entirely that of the employer.

Two logbooks should be maintained for any workshop. One is used for reporting potential hazards so that action can be taken (or if the worst happens, blame apportioned). The other logbook is used to record actual accidents, so that a written account is available of every incident.

Referring to these logs at regular intervals, perhaps at the start of each month, is a valuable way of revising safety policies.

First aid

In every workshop at least one person, and preferably two, should be trained in first aid procedures.

- First aid need not involve elaborate treatment, and most first aid is concerned with minor cuts and burns.
- A fairly basic first aid chest along with good facilities for washing is adequate for at least 90% of incidents.
- Treatments for exposure to toxic materials or fumes call for much more specialized equipment and knowledge.
- It is particularly important never to work alone with potentially toxic materials.
- Training to deal with electric shock should be a first priority of first aid instruction for servicing workshops.
- The power supply must be cut off before the victim is touched, otherwise there may be two victims instead of one.
- Mouth-to-mouth resuscitation, preferably by way of a plastic disposable mouthpiece, should then be applied as soon as the power is off, and continued until the victim is breathing or until an ambulance arrives.
- If it would take too long to find and operate a mains switch, push the victim away from live wires using any insulated materials, such as a dry broom handle, or pull away by gripping (dry) clothing.
- Do not touch the skin of the victim. This endangers the rescuer and also the victim (because of the additional current that will flow, perhaps through the heart).
- Speed is important, because, even if breathing can be restored, the brain can be irreparably damaged after about four minutes.
- Always leave an unconscious victim on his/her side, never on the back or on the face.
- Never try to administer brandy or any other liquid to anyone who is not fully conscious because the risk of choking to death is at least as great as that of electric shock itself.

Once the victim has been removed from the danger of continued shock, a 999 call can be made and mouth-to-mouth resuscitation can be given. Practical experience in this work is essential, and should form an important part of any first aid training. A summary will remind you of the steps, and posters showing the method in use should be displayed in the workshop.

1. Place the victim on his/her back, loosen any clothing around the neck, remove any items from the mouth such as false teeth or chewing gum.
2. Tip the head back by putting one of your hands under the neck and the other on the forehead. This opens the breathing passages.
3. Pinch the nose to avoid air leakage, breathe in deeply, and blow the air out into the victim's lungs. If possible, use an approved mouth-to-mouth adapter to avoid any risk of transferring disease, but never waste time looking for one.
4. Release your mouth and watch the victim breathe out. You may have to assist by pressing on the chest.
5. Repeat at a slow breathing rate until help arrives or the victim can breathe unattended. Do not give up just because the victim is not breathing after a few minutes because these efforts can sustain life and avoid brain damage even if the victim is unconscious for hours.

Electric shock is often accompanied by the symptoms of burning which will also have to be treated, though not so urgently. The workshop telephone should have permanently and prominently placed next to it a list of the numbers of emergency services such as doctors on call, ambulance, fire, chemists, hospital casualty units, police, and any specialized services such as burns and shock units. This list should be typed or printed legibly, maintained up to date and stuck securely to a piece of plywood or hardboard. Part of any safety inspection should deal with checking that this list is updated, well placed, and legible.

- Severe burning must be treated quickly at a hospital.
- First aid can concentrate on cooling the burns and treating the patient for shock.
- Apply cold water to the region of the burn and when the skin has had time to cool, cover with a clean bandage or cloth.
- Never burst blisters or apply ointments, and do not attempt to remove burned clothing because this will often remove skin as well.
- The reaction to burning is often as important as the burn itself, and any rings, bracelets, tight belts and other tight items of wear should be removed in case of swelling.

Try to keep the patient conscious, giving small drinks of cold water, until specialist help arrives.

The treatment of minor wounds is a frequent cause of a call on first aid.

- The most important first step is to ensure that a wound is clean, washing in water if there is any dirt around or in the wound.
- Minor bleeding will often stop of its own accord, or a styptic pencil (alum stick) can be used to make the blood clot.
- More extensive bleeding must be treated by applying pressure and putting on a fairly tight sterile dressing; one dressing can be put over the top of an older one if necessary rather than disturbing a wound.
- Medical help should always be summoned for severe wounding or loss of blood, because an antitetanus injection may be required even if the effects of the wound are not serious.

The effects of chemicals require specialized treatment, but first aid can assist considerably by reducing the exposure time. In electronics servicing work, the risk of swallowing poisonous substances is fairly small, and the main risks are of skin contact with corrosive or poisonous materials and the inhalation of poisonous fumes.

If a corrosive chemical has been swallowed or spilled on the skin, large amounts of water should be used (swallowed or used for washing) to ensure that the material is diluted to an extent that makes it less dangerous.

Common solvents like trichloroethylene and carbon tetrachloride degreasing liquids give off toxic fumes. These liquids should be used (other than in very small quantities) only under extractor hoods or in other well-ventilated situations.

- Never try to identify a solvent by sniffing at a bottle. Bottles should be correctly labelled, preferably with a hazard notice.
- Some solvents, like acetone and amyl acetate, are a serious fire hazard in addition to giving off toxic fumes.

CARE OF EQUIPMENT UNDER REPAIR

While equipment is being serviced, it is the responsibility of the servicing engineers to ensure that the equipment is not damaged. Damage in this respect means mechanical or electrical damage, and though such hazards should be covered by insurance, the customer is unlikely to feel well disposed to any servicing establishment which cannot take better care of equipment that is being serviced.

Mechanical damage covers such items as:

- Damage to either the cabinet or the functioning of the circuitry, or both, caused by dropping the equipment.
- Tool damage, such as burn marks on a cabinet caused by a soldering iron.
- Marks or stains, such as can be caused by a hot coffee cup or by carelessness with solvents.

Such mechanical damage might be due to an unavoidable accident, but only too often it is the result of carelessness, and the ultimate loser, regardless of insurance, will be the staff of the workshop. Never assume that an old piece of equipment is of low value to the customer, or that a customer will not worry about damage. For example, some older models of TV receivers were housed in cabinets that were fine pieces of carpentry, and the owner still uses the equipment because there is nothing manufactured now that can replace it. The fact that such a receiver is 18 years old will not weigh with the customer if it is returned with a prominent burn mark, and there will be little or no chance of making a repair or replacement, either to the cabinet or to your reputation.

Since mechanical damage to the cabinet is the form of damage that is most obvious to the customer, all cabinets should be wrapped both when being handled and in the course of servicing. Bubble-pack is particularly useful for avoiding damage caused by knocks, and this can also protect against marking or staining. Damage from soldering irons can be minimized by using holders for irons, and ensuring that cabinets are kept well away from tools.

One other hazard of this type is ingrained swarf, including fragments of solder caused by shaking a soldering iron. If the workshop is not kept clean, metal swarf can become embedded in the underside of a cabinet. Though this will not be visible to the customer, it can cause severe scratching on the table that the equipment stands on, and the customer is not likely to forget or forgive the damage.

Electrical damage is typically caused by:

- Incorrect use of servicing equipment such as signal generators.
- Carelessness in removing ICs.
- Ignorance of correct operating conditions.
- Replacement of defective components by unsuitable substitutes.
- Operating equipment with incorrect loads (including o/c or s/c load).
- Failure to check power supply components when the initial fault has been caused by a power-supply fault.

One feature that is common to many of these items is incorrect or unavailable information. Equipment should not be serviced if no service sheet is available. Granted that there are some items which are so standardized that no service sheet is necessary, but these items, such as radios and personal stereo players, are the ones that are uneconomical to service in any case. Equipment such as TV receivers, Hi-Fi, personal computers, video recorders, etc. will require information to be available, and that information should be as complete as possible. Service engineers must be aware which components must be replaced only by spares approved by the manufacturers, and why. Substitution should be considered only if the manufacturer no longer supplies spares – remember that some Japanese manufacturers will no longer supply spares after a comparatively short period, as short as 5 years.

RELIABILITY

The reliability of equipment is often thought of as a matter that is entirely for the manufacturer, and of little concern to the service engineer. This is not so, and the service engineer is a vital link in the quest for greater reliability. It is only by keeping in close touch with a manufacturer on faults that have appeared in equipment that improvements in reliability can be carried out.

In particular, some faults seem to turn up over and over again. Eventually these will be logged on service sheets, but only if they have been reported to the manufacturers. Eventually, of course, the requests for spares can alert a manufacturer to problems, but this takes much longer than a report from a service engineer. Such reports need not be too detailed, and some manufacturers provide for the use of Email, or at least a fax number, for reporting problems.

Being aware of probable faults can greatly speed up servicing, and a good service sheet should list known faults that seem to crop up frequently. Close attention to trade magazines will often reveal faults which might be baffling without the benefit of experience, and a file of these hints and tips should be kept. This is relatively easy for a workshop that specializes in one make, but if you have to service a wide range of equipment, such files can become untidy, and it is tempting to use a computer as a way of coping with the mass of information. Sadly this is not always an answer, because someone has to enter the information into the computer, and it is not always easy to enter such items as circuit diagrams unless the workshop is unusually well equipped (with a scanner, for example).

Reliability is usually measured in terms of mean time before failure

(MTBF), which as the name suggests, is the average time that equipment will operate before a failure. This is no place to become involved with the (often difficult) statistics of reliability, but the higher the MTBF the more reliable the equipment. A figure of MTBF of 5 years is no guarantee that a piece of equipment will not fail the first time it is switched on, but it does suggest that if you have sold several dozen pieces of the same equipment, you can expect not to have to service many of them for some time to come.

How many will require servicing in 2 years, 3 years, or 5 years? There is no simple answer, but the manufacturers who quote a figure for MTBF should be able to supply a chart showing the likelihood of a failure occurring after a specified period. As always, dealing with probabilities and chances is an uncertain business, and it can never be applied to small numbers. If you have supplied just one of a particular model, you can forget about MTBF and any other statistics, because statistics are meaningless for small numbers. You might have a 90% chance of surviving to the age of 68, but this does not mean that you will survive walking in front of a moving bus at the age of 25.

Unless you are supplying large numbers of one model of equipment, you should assume that all will require service of some sort in the normal lifetime of electronic equipment, which nowadays is close to 10 years. If the manufacturer's guarantee is for a short period you should expect that the rate of servicing will be higher than for equipment for which a manufacturer expresses more confidence.

OUTSIDE INSTALLATION WORK

Outside installation work, as distinct from outside servicing, is unavoidable, and for consumer electronics often refers to aerials (FM, TV or satellite), down-leads and interconnections. The installation of personal computers, however, also comes into this class, and though aerial work is specialized and will usually be contracted out (and will not be further considered here), the installation of Hi-Fi and computers will often have to be done by the supplier. As always, this should be realistically priced. If the customer suggests that he/she can buy the boxes elsewhere for a lower price, point out by all means that you will supply an installation service at a price that reflects a fair return, and that such installation is an integral part of the price that you quote for equipment.

Outside installation follows the same pattern as was mentioned earlier for outside repairs, but with no option for taking the equipment to the workshop. The customer needs to be consulted about positioning of equipment – the loudspeaker may, after all, clash with the colour of the curtains – but the installer should be able to point out tactfully that a full stereo effect will not be experienced if the loudspeakers are placed close to each other, facing in different directions or even (as I have seen) in different rooms. Similarly, it is up to the installer to show that the monitor of a computer should be placed where the user of the keyboard can see it without needing to lean forward or sideways, that the mouse should come conveniently to hand, and that the flow of air through the main casing is unobstructed.

Though the interconnection of equipment is often so standardized that you can work without manuals, you should be aware of any peculiarities of equipment, limitations to cable lengths (particularly for parallel printers for computers), etc. A useful hint for computers that are used with a large amount of powered peripherals (such as scanner, modem, printer, etc.) is to connect the power take-off socket on the main computer case to a set of distribution sockets, which can be used to supply to peripherals. This allows the user to switch everything on and off together using only the main switch on the computer. A similar scheme can be used for the more elaborate type of Hi-Fi setup.

Never leave an installation without testing it adequately – do not assume that because it sounds good playing a cassette that the CD player will be as good. In some cases, you may need, for example, to alter sensitivity levels at inputs. In addition, never leave an installation without making sure that the customer knows how to operate the equipment and what peculiarities of installation are present. Few users, for example, know how to deal with the connections between a modern VCR and TV receiver, and many assume that video replay still needs the use of a spare channel rather than the AV option.

CHAPTER 2

THE WORKSHOP

NEW WORKSHOP SUPPORT TOOLS

TV and monitor pattern generators

Generally the modern workshop is well provided with robust and weighty instruments for the setting up and checking of the picture geometry and colour scales, while the field service engineer has in the past had to struggle along with the minimum of support equipment. However, this situation has now changed for the better. Using one of the hand held devices that are now well established, the field engineer can now be equipped with an economical instrument that is robust enough to be carried around in a tool box and still be accurate enough to provide a set of test signals to broadcast standards.

The Teletest 2 units (courtesy of Nick Rose of OZAN, Wimbourne, Dorset) are shown in Figure 2.1. These are either battery operated or powered via a mains adapter plug unit, and form part of the set of pattern generators designed to cover most of the needs of service engineers ranging from those involved in professional broadcasting to the consumer electronics industry. With costs ranging from around £100 to £200 these are economically priced to allow each engineer to be equipped with his or her own personal instrument.

The units provide outputs of video or audio signals at baseband or UHF, for PAL I or PAL G systems, with signal parameters to CCIR Broadcast Specification standards. The video output at UHF is nominally tuned to channel 36, but this can be varied over the range of channels 32 to 42. These are therefore ideally suited for those engineers involved in the retuning of VCRs and receivers for Channel 5 television. The 1 kHz audio output is clean enough and free from

Figure 2.1 The Teletest 2 unit.

distortion to allow it to be used for the most demanding of test purposes. In particular, this signal is useful for locating faults and tracing the cabling associated with the home cinema surround sound systems.

All units provide at least six fully interlaced patterns. The red, white and black synchronized rasters are invaluable for purity testing and making black/white level balance adjustments. The rock steady crosshatch pattern with border castellations makes it easy to perform convergence and picture set-up operations. In a similar way, the two grey scale and rainbow bar patterns proved to be most useful for the exercise of setting brightness, contrast and colour tracking.

Bench power supplies

These are particularly useful for powering camcorders, car radios and even VCR mechanisms. They can also be pressed into service as battery chargers or to provide biasing for servos and tuner units. The typical requirement is for a voltage range of about 3–30 volts with output currents up to about 25 amps. The circuit should be well stabilized and equipped with both voltage and current meters. Digital read out can be valuable when attempting to locate intermittent faults in one system while working on another.

Portable appliance testers (PAT)

It is important to be able to test and record the leakage and earth loop impedances and insulation between line, neutral and earth after repairs. This will ensure that the work carried out meets the standard now required by law.

WORKSHOP SUPPORT DATABASES

Engineers servicing domestic electronic equipment are generally well provided with circuit diagrams and data sheets, but to produce a thoroughly reliable and economical repair that will generate customer loyalty, they need additional skills and support. Not only is it necessary to understand how the system works, there is also a need to have a knowledge of the system's historical reliability and its particular points of failure. In days gone by when systems were constructed largely from discrete components, many of the failures occurred in a regular manner, giving rise to the *stock faults*. Many service departments found these to be a source of good business that created a sound reputation for doing a good job.

With today's extensive use of dedicated ICs the system reliability has improved very considerably. However, stock faults still occur but repeat very much less frequently, making a good memory an additional requirement for the service engineer. Fortunately, there are now a number of computer system databases available that have been designed to aid service personnel in the repair of equipment. While these can provide almost instant access to many of the stock faults, it must be emphasized that these should be used in conjunction with the manufacturer's circuit diagrams and data sheets. Of these databases, two have been tested and found to be invaluable to the busy workshop.

The first one of these was provided by SoftCopy Ltd, Electronic Publishers, Cheltenham, Gloucester, GL53 0NU. This has been produced in co-operation with the journal *Television* published by Reed Business Publications and is updated annually. The program has been designed to run on a relatively low level IBM compatible personal

```
 [■]              TELEVISION Index and Directory 1996

Advertisers Index                  | WELCOME TO THE FOURTH ISSUE OF
  DX-TV                            | the TELEVISION Index and Directory
■ Fault Finding: CD Player Casebook | on disk.  It spans eight years of
              : CD Player Notes     | TELEVISION magazine from volume 38
              : Camcorder          | 1988 to volume 45 1995.
              : Camcorder Notes     |
              : Satellite TV        | Improvements this year include
              : Satellite TV Notes  | the ability to input your own
              : Service Bureau      | fault reports, and proper printing
              : TV Fault Finding    | of reports from the Fault Report
              : TV Fault Notes      | Disks.  Help has been improved
              : Test Case           | by making the manual available
              : VCR Clinic          | on-line at any time by pressing the
              : VCR Fault Notes      | F1 key.
  General Notes                    |
  Information                       | Fault Report Disks are available
  International TV Standards         | for volumes 41-45.  Earlier volumes
  Leaders                           | will be issued in 1996.  Watch our
  Letters                           | advert in TELEVISION.
  Microcomputers                    | (c) SoftCopy Limited 1995

  F1-Help  F2-Search                       F10-Exit        3/41
```

Figure 2.2 Introduction page for *Television* database.

computer, contains about 7000 different entries and is very easy to use. There is additional space within the system memory to incorporate the results of personal experience. The extent of the information that is provided is indicated by the examples of printouts included here.

The database covers faults on CD players, camcorders, satellite TV systems, VCR machines as well as television receivers that have been published in the journal back as far as 1988. In addition, there is a directory reference to many of the published articles that date back as far as 1986. These provide good background reading to obtain an understanding of the operations of many of the popular items of equipment. It is thus a tool that should find a slot in any repair workshop. Because of its relatively low processing needs, this database could also be installed on a low-cost laptop computer for use by the field servicing engineers.

By comparison, the database provided for testing by EURAS International Ltd, Keynsham, Bristol, BS18 2BR, is supplied on a CD-ROM which to some extent increases processing needs. However, this is updated three times annually on a subscription basis and covers the same wide range of domestic entertainment equipment. The faultfinding and repair hints represent the knowledge gleaned from manufacturers, dealers and repair centres, and covers about 250 000 entries from more than 400 manufacturers. Unusually, the database information is covered in most European languages. Again the system memory provides space to record personal knowledge gained over a period of time. Because of the greater available memory space on the CD-ROM, the publishers can provide the appropriate circuit diagram segments with each repair hint. Samples of the print-outs obtained show the extent of the helpful information provided.

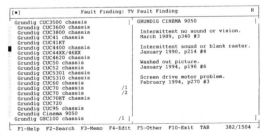

```
 [■]         Fault Finding: TV Fault Finding              R

Grundig CUC3500 chassis        | GRUNDIG CINEMA 9050
  Grundig CUC3600 chassis       |
  Grundig CUC3800 chassis       |   Intermittent no sound or vision.
  Grundig CUC41 chassis         |   March 1989, p340 #3
  Grundig CUC41KT               |
■ Grundig CUC4400 chassis        |   Intermittent sound or blank raster.
  Grundig CUC44XX/46XX          |   January 1990, p214 #4
  Grundig CUC4620 chassis       |
  Grundig CUC50 chassis         |   Washed out picture.
  Grundig CUC52 chassis         |   January 1994, p196 #6
  Grundig CUC5301 chassis       |
  Grundig CUC5310 chassis       |   Screen drive motor problem.
  Grundig CUC60 chassis         |   February 1994, p270 #3
  Grundig CUC70 chassis      /1 |
  Grundig CUC70 chassis      /2 |
  Grundig CUC70KT chassis       |
  Grundig CUC720                |
  Grundig CUC95 chassis         |
  Grundig Cinema 9050           |
  Grundig GSC100 chassis     /1 |

  F1-Help  F2-Search  F3-Memo  F4-Edit  F5-Other  F10-Exit  TAB   382/1504
```

Figure 2.3 Faultfinding page of *Television* database.

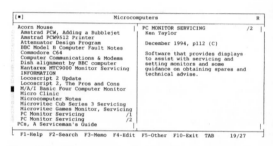

Figure 2.4 Microcomputer data sample, *Television* database.

BASIC UNITS, FORMULAE, AND PREFIXES

The basic units of electronics are the electrical units, of which the most familiar are the volt, ampere (amp), ohm, farad and hertz. To this can be added the henry, used for mutual or self-inductance, the joule and the watt, and the basic length unit, the metre.

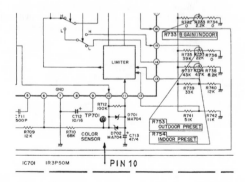

Figure 2.5 Euras camcorder example, with magnified circuit.

Electronics, both in theory and in practice, often requires you to work with numbers that are either very large or very small. Quantities greater than 100 or less than 0.01 are usually expressed in the standard form of $A \times 10^n$, where A is a number, called the *mantissa*, which is less than 10, and n is a whole number called the *exponent*. A positive value of n means that the number is greater than unity, a negative value of n means that the number is less than unity.

For example, we write 120 000 as 1.2×10^5 (or 1.2E5) because this is shorter and because there is less chance of error in copying the zeros. The zeros in this number are used only to indicate the powers of ten; they mean that the number is 1.2 times one hundred thousand. By contrast, the zero in the number 102 means that this number contains no tens, though it consists of one hundred units plus two units. Taking another example, 0.000 5 can be written as 5×10^{-4} because here again, the zeros convey the power of ten and are not part of the significant figures of the number.

Zeros are called *non-significant* if they follow the significant numbers or if they precede them. The zeros in 23 000 and in 0.0012 are not significant, but the zeros in 2003 and 0.1005 *are* significant. When we write numbers in this abbreviated form, called *scientific form*, we are separating the non-significant zeros.

How many significant figures do we need in any number? As an example, the number 12 003 is obviously more than 12 002 and less than 12 004, but is this difference important? For example, if you were measuring a distance in millimetres as 12 003 mm, could you be confident that your measurement was precise to 1 millimetre in this amount? You could certainly not be so precise if you used any form of tape measure (it would stretch by more than 1 mm if you pulled it a little bit tighter), and there are not many ordinary ways of measuring distance that could justify that final figure '2'. We might as well call this number 12 000 mm, so that there are really only two significant figures, the figures that matter. A lot of measurements do not justify more than three significant figures, and some cannot really be more precise than two significant figures.

This ties up with the standard form for numbers, because it allows us to be more economical with figures. We don't, for example, often need to convert a number like 1 217 563 into $1.217\,563 \times 10^6$, because it would be pointless unless we have really measured this amount correct to one part in a million. It would be more realistic to write the number as 1.217×10^6, or even as 1.22×10^6, rounding it off to three significant figures. The one item in electronics work that quite certainly needs to be expressed with a large number of significant figures is the frequency of a crystal. For example, you can think of the TV colour subcarrier frequency as 4.43 MHz, but you need to remember that it is actually 4.433 618 75 MHz – it has to be as precise as this.

Conversions to/from standard form

To convert a number into standard form, shift the decimal place until the portion on the left-hand side of the decimal point is between 1 and 10, and count the number of places that the point has been moved. This is the value of n. If the decimal point has had to be shifted to the left the sign of n is positive; if the decimal point had to be shifted left the sign of n is negative. For example:

$$1\,200 \text{ is } 1.2 \times 10^3 \quad \text{and} \quad 0.001\,2 \text{ is } 1.2 \times 10^{-3}$$

To convert numbers back from standard form, shift the decimal point n figures to the right if n is positive or to the left if n is negative.

For example:

$$5.6 \times 10^{-4} = 0.000\ 56 \quad \text{and} \quad 6.8 \times 10^5 = 680\ 000$$

Note in these examples that a space has been used instead of the more familiar comma for separating groups of three digits (thousands and thousandths). This is recommended engineering practice and it avoids confusion caused by the use, in other languages, of a comma as a decimal point.

Numbers in standard form can be entered into a calculator by using the EXP or E key where you would write the ×10 part on paper. For example 1.2×10^6 would be keyed in as 1.2E6, and 5.4×10^{-7} would be entered as 5.4E-7 (you would press the 7 key and then the ± key). See later for the use of calculators.

Prefixes

Where formulae are to be worked out, numbers in standard form are normally used, but for writing component values it is more convenient to use the prefixes shown in Table 2.1. This allows a name to be used in place of the $\times 10^n$ portion of a number, and you need to know which name refers to which multiple of ten.

- The prefixes have been chosen so that values can be written without using small fractions or large numbers.
- Using these standard prefixes avoids the need to have to write powers of ten, and it allows quantities to be written very concisely. For example, 24 000 volts can be written as 24 kV and 0.000 000 1 farads as 0.1 μF or 100 nF.
- A few formulae can be written in a form that uses the most likely prefixed units, for example, in mA and kΩ rather than in V and Ω, but you then need to remember what units are required.
- Note that 1 000 pF = 1 nF; 1 000 nF = 1 μF and so on.
- In *computing*, the K (rather than k) symbol is used to mean 1 024 rather than 1 000 and M means 1 048 576 – these quantities are the nearest exact powers of two.

Table 2.1 The names that are used for powers of ten. The **bold** print indicates the most common prefixes. The centi-, deci-, deca- and hecto- prefixes are not normally used for electrical quantities

Prefix	Abbreviation	Power of Ten	Multiplier
tera	T	10^{12}	1 000 000 000 000
giga	**G**	10^9	**1 000 000 000**
mega	**M**	10^6	**1 000 000**
kilo	**k**	10^3	**1 000**
hecto	h	10^2	100
deca	da	10^1	10
deci	d	10^{-1}	1/10
centi	c	10^{-2}	1/100
milli	**m**	10^{-3}	**1/1 000**
micro	**μ**	10^{-6}	**1/1 000 000**
nano	**n**	10^{-9}	**1/1 000 000 000**
pico	p	10^{-12}	1/1 000 000 000 000
femto	f	10^{-15}	1/1 000 000 000 000 000

Ratios

Many of the quantities used in electronics measurements are ratios, such as the ratio of the current (i_c) flowing in the collector circuit of a transistor to the current (i_b) flowing in its base circuit, equal to h_{fe}.

A ratio consists of one number divided by another, and it can be expressed:

(a) As a common fraction, such as 2/25.
(b) As a decimal fraction, such as 0.47.
(c) As a percentage, such as 12% (which is in fact a short way of expressing the fraction 12/100)

- To convert a decimal ratio into a percentage, simply shift the decimal point two places to the right, so that 0.47 becomes 47%. If there are empty places, fill them with zeros, so that 0.4 becomes 40%.
- To convert a percentage to a decimal ratio, imagine a decimal point where the % sign was, and then shift this point two places to the left, so that 12% becomes 0.12. Once again, empty places are filled with zeros, so that 8% becomes 0.08.
- To convert a decimal to a fraction, write as the numerator (top portion) of the fraction the figures that follow the decimal point. Write as the denominator (the lower portion) a '1' in place of the decimal point, and as many zeros as there are figures following the point. For example, 0.156 is the fraction 156/1000 and 0.07 is the fraction 7/100.
- To convert a fraction to a decimal, divide the numerator by the denominator (easiest using a calculator) until there is no remainder. For example, the fraction 3/14 is converted in the following steps:
  - 1 divided by 14 cannot be done, write the 0. portion of the decimal and place a zero following the '3'. Result so far is 0
  - 30 divided by 14 is 2 and 2 remaining. Write down the 2 following the decimal point and place a zero following the remainder of 2. Result so far is 0.2
  - 20 divided by 14 is 1 with 6 remaining. Write the '1' as part of the decimal, and place a zero following the '6'. Result so far is 0.21
  - 60 divided by 14 is 4 and 4 remaining. Add the '4' to the decimal and place a zero following the '4'. Result so far is 0.214
  - continue until the division leaves no remainder or until there are as many decimal places as you need to work to. In this example, the calculator gives 0.214 285 7, with no indication that the process will end.

Averages and tolerances

The average value of a set of values is found by adding up all the values in the set and then dividing by the number of items comprising the set. Suppose that a set of 19 resistors consists of the following values: one 7R, two 8R, three 9R, four 10R, four 11R, three 12R, and two 13R.

The average value of the set is found as follows:

$$\frac{7+8+8+9+9+9+10+10+10+10+11+11+11+11+12+12+12+13+13}{19} = \frac{196}{19}$$

This divides out to 10.32, so that the average value of the set is 10.32 ohms or 10R32.

⇒ Note that such an average value is often not 'real', in the sense that there is in the set no actual resistor having the average value of 10R32. It is like saying that the average family size in the UK today is 2.2 children. This might indeed be a quite truthful statement but you will seldom in practice meet a family containing two children and 0.2 of a third one.

In a manufacturing process, the average value becomes a target value. When the value of a component turns out to be more or less than the target value, it is said to have a *tolerance*. For example, in a box containing 10R resistors, an 'odd one out' resistor which has a measured value of 12R could be said to have 'a tolerance of 2.0R'.

- Tolerances are in practice more usually expressed as percentages. Thus 2R/10R expressed as a percentage becomes $2/10 \times 100\% = 20\%$, so that the tolerance of the odd resistor is said to be within 20% of the target value.
- A tolerance of $\pm 20\%$ means that some samples in a given batch may be as much as 20% high (+) and others 20% low (−) compared to the target or average value of the batch as a whole.

Though it should be a familiar point, it is possibly worth pointing out that components are separated out by the manufacturers, so that it is fruitless to sort through a box of 1K 20% resistors looking for one that has an exact 1K value. A set of 20% resistors will consist of resistors that have values between 10% and 20% more or less than 1K, certainly not any that are 1K.

Equations

Equations are a shorthand way of writing relationships between physical quantities. The purpose of an equation is to make it possible to work out the size of a quantity, given the values of the other quantities that it depends on. The equation is given in terms of letters rather than numbers, because this allows you to put in the numbers that you want to use.

For example, in the equation $V = RI$, the quantity represented by V is found by multiplying the quantity represented by I by the quantity represented by R. If $I = 5$ and $R = 20$, then V must be 100. In all equations multiplication is normally indicated by the use of a dot, such as $f.C$, or simply by close printing such as $2\pi f C$.

The units that must be used with such formulae are also shown and *must* be adhered to – if no units are quoted then fundamental units (amp, ohm, volt, etc.) are implied. For example, the equation:

$$X = 1/2\pi f C$$

is used to find the reactance (X) of a capacitor in ohms, using C in farads and f in hertz. If the equation is to be used with values given in μF and kHz then values of 0.1μF and 15 kHz are entered as 0.1×10^{-6} and 15×10^3 respectively. Alternatively, the equation can be written as $X = 1/2\pi f C$ MΩ using values of f in kHz and C in nF, and with the answer in megohms, millions of ohms. In this example, the equation looks unchanged, and considerable care is needed if you want to use units other than the fundamental units.

Where brackets are used in an equation, the quantities within the brackets must be worked out first, and where there are brackets within brackets, the portion of the equation in the innermost brackets must be worked out first, followed by the material in the outer brackets. For example:

$$2(3 + 5) \text{ is } 2 \times 8 = 16 \quad \text{and} \quad 2 + (3 \times 5) \text{ is } 2 + 15 = 17$$

Apart from brackets, the normal order of working out is to carry out multiplications and divisions first followed by additions and subtractions.

Transposition

Any equation can be expressed in more than one way. For example, in the familiar $V = RI$ equation, the quantity V is the subject of the equation. If the equation is rewritten so that the *subject* is changed, for example as $R = V/I$, then we have *transposed* or *changed the subject* of the equation. Sometimes an equation is given in all the forms that you might want to use, but if it is not, you need to know how to transpose the equation so that its subject is the unknown quantity whose value you want to find.

Transposing is simple provided that the essential rule is remembered:

- An equation is not altered by carrying out identical operations on each side.

For example:

$$Y = \frac{5aX + b}{C}$$

is an equation that allows you to find Y when the values of a, X, b and C are known. This equation can be transposed so that it can be used to find the value of X when the other quantities of Y, a, b, and C are known. The procedure is to keep changing both sides so that X is left isolated. The steps are as follows:

(a) multiply both sides by C result is $CY = 5aX + b$

    – remember that $\dfrac{C(5aX + b)}{C}$ can be followed by cancelling C since multiplying $(5aX + b)$ by C and then dividing it by C leaves it unchanged at $5aX + b$.

(b) subtract b from both sides   result is $CY - b = 5aX$

(c) divide both sides by 5a   result is $\dfrac{CY - b}{5a} = X$

so that the equation has now become $X = \dfrac{CY - b}{5a}$ which is the transposition we required.

Commonly encountered equations

The ohm set:	$V = RI$; $R = V/I$ and $I = V/R$
Resistors in series	$R_{\text{total}} = R_1 + R_2 + R_3 + \ldots$
Resistors in parallel	$1/R_{\text{total}} = 1/R_1 + 1/R_2 + 1/R_3 + \ldots$
Capacitors in series	$1/C_{\text{total}} = 1/C_1 + 1/C_2 + 1/C_3 + \ldots$
Capacitors in parallel	$C_{\text{total}} = C_1 + C_2 + C_3 + \ldots$
Reactance of capacitor	$Xc = 1/2\pi fC$
Reactance of inductor	$XL = 2\pi fL$
Resonance (series L and C)	$f = 1/(2\pi LC)$
Tr. voltage gain, common-emitter	$G = 40 V_{\text{bias}}$

Inductance of a coil of n turns wound on a core of specific inductance A_L is:

$$L = n^2 A_L$$

Temp coefficient formulae:

For a metal:

$$R_2 = R_1 (1 + \alpha\theta)$$

where R_1 is resistance at lower temperature, R_2 is resistance at higher temperature, θ is the temperature difference, and α is the coefficient of resistivity (or resistance).

For a thermistor, use:

$$R_2 = R_1 \exp\left(\frac{B}{\theta_2} - \frac{B}{\theta_1}\right)$$

where B is the thermistor constant, θ_2 and θ_1 are the two temperatures in Kelvin units (= °C + 273), and R_1 and R_2 are the corresponding resistance values. The quantity in brackets should be worked out first, and the value used as the starting figure for the e^x function of a calculator.

GREEK ALPHABET

Table 2.2 shows both upper (capital) and lower-case letters, along with pronunciation. Note that the pronunciations are for classical Greek, and some, such as β, are different in modern Greek. The letter Σ (sigma) in classical Greek has two lower-case forms written usually as σ, but as s at the end of a word. Only the σ form is illustrated here.

Table 2.2 Greek letters and their names

Greek		Name	Greek		Name	Greek		Name
A	α	alpha	I	ι	iota	P	ρ	rho
B	β	beta	K	κ	kappa	Σ	σ	sigma
Γ	γ	gamma	Λ	λ	lambda	T	τ	tau
Δ	δ	delta	M	μ	mu	Y	υ	upsilon
E	ε	epsilon	N	ν	nu	Φ	φ	phi
Z	ζ	zeta	Ξ	ξ	xi	X	χ	chi
H	η	eta	O	o	omicron	Ψ	ψ	psi
Θ	θ	theta	Π	π	pi	Ω	ω	omega

TERMINOLOGY

When quantities such as 3.5 k or 6.8 V are written on sheets that will be used in workshop conditions, there is a risk that the decimal point will be misread. A photocopied sheet, for example, may not reproduce the points, or a piece of dirt may appear to be a point. The method laid down in BS 1852 avoids this problem, and also avoids the problems that are caused by the reversed use of points and commas as used on in other EEC countries. The system was primarily used for resistor values, but has been extended to the stabilized voltages of Zener diodes and cells.

In the BS 1852 system, the resistance unit letter replaces the decimal point, using the letters R, K and M in the usual sense. For example, a 0.33 Ω resistor will be written as R33 (or 0R33), and a 4.7 Ω resistor as 4R7. Values such as 22K and 33K are unchanged in this system, though 6.8 K is written as 6K8. A 2.2 MΩ resistor appears as 2M2 and so on.

Where tolerance is important, a letter following the value can be used to express the tolerance value as follows:

M = 20%   K = 10%   J = 5%   G = 2%   F = 1%

For Zener diodes, the letter V is used in place of the decimal point, so that values of 3V3, 4V7, 6V8, etc. can be written.

MEASUREMENTS USING dBs

For most purposes, the gain of amplifiers is quoted in decibels (dB). This is a logarithmic scale which is better suited for expressing gain for several reasons that are mostly tied up with the way that gain affects our senses. Doubling the gain of an audio amplifier, for example, makes the sound seem louder, but certainly not twice as loud. This was recognized by the telephone pioneer Alexander Graham Bell, and so the unit of loudness was named the bel in recognition of this. The bel unit is too large, however, and one tenth of a bel, a decibel, is more useful.

The definition of a bel is in terms of power ratio, equal to the logarithm of the ratio of two powers, usually output power and input power, and the decibel is one tenth of a bel. The decibel value of a ratio is therefore:

$$10 \log \frac{P_1}{P_2}$$

where P1 and P2 are the two power levels being compared, such as the output and input power, respectively, of an amplifier.

This is the only precisely correct definition of the decibel ratio, but for comparing voltage gains, the formula

$$20 \log \frac{V_1}{V_2} \text{ is used}$$

The idea behind this is that power is proportional to voltage squared, and in logarithmic terms, squaring is carried out by multiplying a logarithm by 2, making the numerical factor 20 in place of the 10 that is used for the power decibel. This form of decibel calculation is very common, but it is strictly speaking applicable only when the voltage levels are measured across the same impedance levels, which is a condition that is very seldom fulfilled.

GRAPHS

Graphs are a very widely used method for presenting information about electronic components, systems and circuits. It is therefore easy to understand how to interpret the information they so conveniently present.

The purpose of a graph is to take the place of a detailed table of measurements in showing at a glance how two quantities are related to one another in differing circumstances. The table of voltages and currents in Table 2.3, for example, lists the results of a series of measurements which have been taken on a given circuit.

Try to guess quickly from this table alone the value of voltage that would result in a current flow of, say, 0.2 A. Then see how much quicker and more accurate your estimate would have been if the same information had been shown in the form of the graph which forms Figure 2.6. In this graph, the relationship between the two sets of data is at once obvious, and the closely approximate values of all current flows and their corresponding voltages can be read off from the graph scales at a glance.

To plot such a graph, two scales are needed. One is drawn along

Table 2.3 A set of measurements which are not easy to interpret by themselves

V in volts	I in amps
0.85	0.07
1.9	0.15
3.0	0.24
3.9	0.31
5.4	0.43

what is called the *X-axis* of the graph. It is the direction which is horizontal when the book you are reading is held upright. This X-axis has marks equally spaced along its length to represent equal steps of (in this particular case) current values. The other axis, called the *Y-axis*, is drawn vertically, at 90° to the X-axis, and in this case is used to plot values of voltage. Equal spaces along this scale thus represent equal steps of voltage.

To plot the graph line itself, in this example, a pair of corresponding values of V (volts) and I (amps) is chosen, and the values of each are located on the appropriate axis. A light line is then drawn vertically upwards (and so parallel to the Y-axis) from the location of the selected current value on the X-axis, and another light line is drawn horizontally (and so parallel to the X-axis) from the location of the selected voltage value on the Y-axis. Where these two lines meet, a lightly pencilled dot is made to form the first point of the graph, its position representing a pair of values, one of current and one of voltage (see Figure 2.7).

Other pairs of values are then similarly located and plotted to form a set of dots. These are then joined by a firmer line, drawn as smoothly and regularly as possible, and so a graph line is produced.

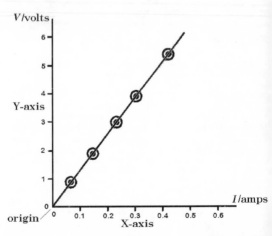

Figure 2.6 A typical linear graph.

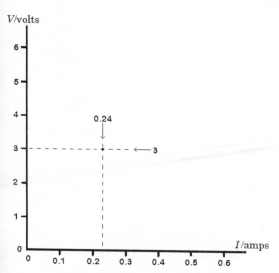

Figure 2.7 Plotting a point from a table of values.

When the dots joined in this way produce a straight line, the graph (and therefore the relationship between the two values being plotted against one another) is said to be *linear*. A linear graph is of great importance because:

(a) It can be extended in either direction simply by using a ruler.
(b) It indicates that one quantity is directly proportional to the other.

The example given in Figure 2.7 is a *linear graph* showing that the voltage across a resistor is proportional to the current flow through the same resistor. This relationship is expressed mathematically in the important equation that we call Ohm's law.

⇒ Note how the axes are labelled with a symbol, oblique stroke and unit quantity. For example. *V*/volts is to be read as '*V*, in units of volts', and *I*/amps as '1, in units of amperes'.

The point 0,0 means the point at which both $X = 0$ and $Y = 0$ and is called the *origin* of the graph. The graph line in Figure 2.6 passes right through this point, indicating as it does that there is no voltage across a resistor when no current flows through it.

If, however, a graph were plotted of the voltages across a rechargeable cell against the current passing through the cell during the charging period (see Figure 2.8), the graph line would again be straight but it would not pass through the origin, because a voltage undoubtedly exists across the cell even when no charging current is flowing.

A graph like this is said to have an *intercept*. The intercept, in this example, is on the Y-axis, which means that there is a value of *Y* when $X = 0$. The intercept is this value of *Y*. In the example, the intercept value is the no-load voltage of the cell.

Another quantity that can be read off a linear graph is its *slope*. The slope of a linear graph is found by taking two points, well spaced from one another, on the graph line, and finding their *X* and *Y* values.

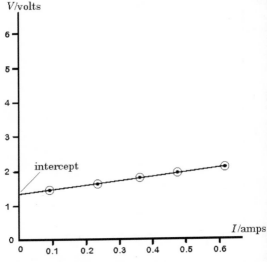

Figure 2.8 A graph with an intercept.

In Figure 2.9 the pairs of points taken are X_1; Y_1, and X_2; Y_2, and their values are as shown on the figure. The lower value of Y_1 is then subtracted from the higher value Y_2, to give $Y_2 - Y_1$, and the X values are subtracted in the same way, to give $X_2 - X_1$. The ratio $\dfrac{Y_2 - Y_1}{X_2 - X_1}$ is the slope of the graph. In Figure 2.9 this slope has a value of 7.88.

⇒ Note that the values of X and Y are always read from the scales marked along each axis, *never* from any measurement of distance made with a ruler.

Suppose, now, that the value of the slope (m) and of the intercept (c) are known for any linear graph. The value of Y for any value of X can then be calculated from the formula:

$$Y = mX + c$$

Example: A graph has a slope value of 5 and an intercept of 2. What is the Y value corresponding to $X = 3$?
Solution: Substitute the data in the equation: $Y = mX + c$. Then $Y = (5 \times 3) + 2 = 15 + 2 = 17$

⇒ Note that in the working out of this example the multiplication was carried out before the addition. It is a general and very important rule that in working out complex equations, all multiplications and divisions are carried out before any additions or subtractions. In olden days, this was memorized in the form MDAS (My Dear Aunt Sally), meaning multiplication,. division, then addition, subtraction.

Non-linear graphs

Not all graphs are linear. For example, a graph of current plotted against voltage in a component called a thermistor would take the

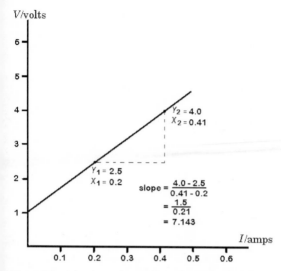

Figure 2.9 The slope of a graph.

shape shown in Figure 2.10. Such graphs have no single value of slope. They are useful for finding instantaneous values of I and V which lie within the range of the graph line, but the graph line itself cannot be extended except by guesswork.

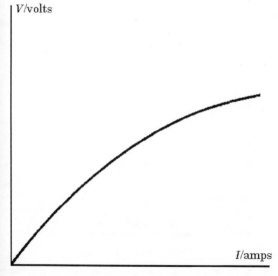

Figure 2.10 A typical non-linear graph.

Sometimes a straight-line graph is obtained by using scales which are not themselves linear. A *non-linear scale* is one in which equal distances along the scale do not represent equal amounts of the quantity being plotted. The graph in Figure 2.11, for instance, uses a logarithmic scale on which each equal step of distance represents a tenfold increase of the quantity being plotted. Logarithmic scales are often used for plotting the frequency of electronic signals, because a very large range of values can then be plotted on a single graph.

One disadvantage of a logarithmic graph is that it is never easy to locate intermediate values lying between the printed values on the axes. For example, the point representing a value of 50 along the X-axis in Figure 2.11 is not midway between 10 and 100 nor midway between 1 and 100. Another disadvantage is that the value of the slope of a logarithmic graph is of little use even if the graph happens to be a straight line.

Another shape, the inverse law, is found when one quantity is related to the inverse of another. For example, the inverse of X is $1/X$, so that a graph of $Y = 1/X$ would take the shape illustrated in Figure 2.12(a) if we plotted X and Y values, but would provide a straight line if we plotted $1/X$ and Y values.

Nomograms

A nomogram is a form of graph which was very popular in the days before calculators. A nomogram consists typically of three (or more) vertical scales, each labelled with the name of a quantity. To use the nomogram, you select a value of one quantity on its scale, and the value of a second quantity on its scale. You then place a ruler or draw a straight line between these points and find where this line cuts the

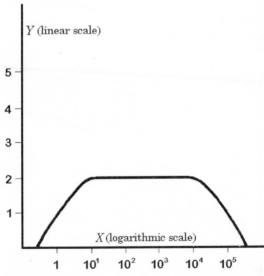

Figure 2.11   A graph with one linear scale and one logarithmic scale.

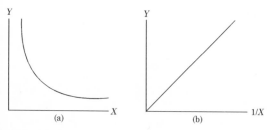

Figure 2.12 An inverse-law graph.

third axis, to give the value of the third quantity. When a nomogram contains more than three axes, you find a value of a third axis as described above, and then use this value combined with a point to another axis to get a value of the final axis.

A nomogram can be a very fast way of obtaining an approximate solution, but like any other system that depends on reading numbers from a scale, its accuracy depends heavily on how precisely the scale can be printed and read. Because of this and the difficulty of constructing nomograms, they have largely fallen out of use. Figure 2.13 shows a simplified example of a nomogram that is used to find the attenuation of signal when a directional aerial is used. In this example, only a small part of the scale is illustrated, and the dotted line shows that at a distance of 30 miles and a frequency of 5 GHz, the attenuation is 140 dB.

SCIENTIFIC CALCULATORS

In the past, tables, graphs and nomograms have been used to help in solving complex calculations, or calculations involving numbers containing many figures. The use of electronic calculators has nowadays made such tasks more or less redundant.

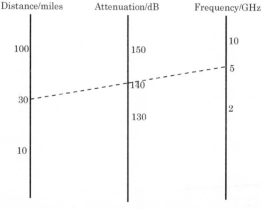

Figure 2.13 An example of a nomogram.

To solve the type of calculations encountered in electronics, a scientific calculator, preferably a type that incorporates all or most of the keys shown in Figure 2.14, is a highly convenient tool. Such a calculator, which need not be expensive, replaces older methods such as sets of tables or the use of slide rules.

The following hints on how to use a scientific calculator assume that the calculator in question is of the 'algebraic' type, rather than being one of those rare models which makes use of the old system of Reverse Polish Notation (RPN).

- For multiplication, division, addition and subtraction, key in the numbers and symbols exactly as in the examples below. The answers will in all cases appear on the display.
- When a number is written in the form of figures having a power of ten (a system usually called 'scientific notation'), use the EXP or EE key as shown below:
- For operations which use the top three rows of keys in Figure 2.14,

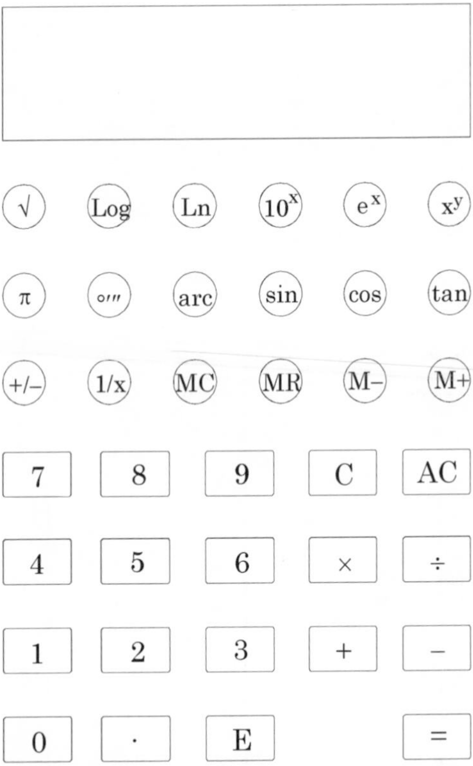

Figure 2.14 A typical scientific calculator keyboard layout, such as that of the Casio calculators.

key in the number first, then the symbol. The answer will appear when the symbol key is pressed. There is no need to press the = key at all.

Examples: To find $\sqrt{5}$ (the square root of 5), use keys 5 and then $\sqrt{}$. The answer appears as 2.236.

To find log 18 (the logarithm of 18), press keys 1, 8, and log. The answer appears as 1.2552.

To find 1/27 (the inverse of 27), press keys 2, 7, and Inv. The answer appears as 0.037.

To find sin 27° (the sine of 27 degrees), key 2, 7, and sin. The answer appears as 0.4539.
Note: The answers given above have been rounded off.

● When problems involving angles need to be solved, you normally need to ensure that a switch on the calculator is set to the correct units for measuring angles – the choice is usually degrees, radians or grads. In all the problems set in this series, the units in which angles are measured will be degrees and the calculator should be set to use degrees for angles.

FAULT-TRACING TECHNIQUES

Basic faultfinding in both analogue and digital systems follows similar principles, that of providing a standard input and tracing the output. A source is required to inject suitable signals into the input and the signal processing is then monitored as it passes through the system on a stage-by-stage basis. For analogue systems a suitable input source is a signal generator, while an oscilloscope can be used as a monitor. For digital systems this end-to-end technique can be carried out using a logic pulser to provide the inputs while the processing is monitored with a logic probe. A technique that is often used with speed advantage is known as the 'half split method'. Here the system is divided into two sections and the end-to-end technique used to find the faulty half. This process is repeated continually until the faulty stage is identified.

Many faults can be traced to overheating, and it is not easy to monitor temperatures inside a cabinet. The solution is to use temperature-sensitive labels which can be stuck at various places inside a cabinet and which will discolour when the temperature is above the value for which the labels are rated.

For more detailed fault-tracing, attention should always be paid to electrolytic capacitors. When electrolytics are operated at high temperatures the effect can be either to dry out the electrolyte or to cause leaks. Drying out will cause a large reduction in the capacitance figure along with lower leakage current; leakage will cause corrosion and eventually an open-circuit in the PCB tracks. See later for notes on cleaning PCBs that have been damaged by leaking electrolytics.

Another electrolytic problem arises from operation with too low a DC bias voltage. This reduces the thickness of the dielectric and will cause a rise in capacitance value initially, followed by a short-circuit failure when the dielectric fails.

Replacement electrolytic capacitors should be of the same (or higher) temperature rating as a component that has failed. TV, VCR and satellite receivers often specify electrolytics capable of prolonged operation at 105°C, and lower-rated capacitors must not be substituted.

TOOLS AND THEIR CORRECT USE

Electronics servicing and minor constructional work requires a small number of tools, but these need to be used correctly and kept in good condition. Correct use of tools is as important from a safety point of view as it is for carrying out the work correctly. You are required as part of the practical test for the C & G 224 course to identify a standard list of tools and state which tools will be required for a given task. The standard list is:

Screwdrivers	Drills and drilling machines	Files
Wire cutters	Pliers	Wire strippers
Soldering iron	Desoldering tools	Hacksaws
Crimping tools	Marking out tools	

and you should know how these tools can be maintained in good condition.

A set of screwdrivers will be needed, ranging in size from the watchmaker's type (1 mm blade) up to a large bladed type for unfastening panels from rack assemblies. Insulated handles are essential, and all-metal screwdrivers should *never* be used for electronics work because of the risk of shorting, since the smaller screwdrivers will be used extensively for adjusting trimmers with voltages applied to the circuit. The blades may be sharpened carefully to keep the original shape indicated in Figure 2.15. Note that cross-point screwdrivers cannot be reground or sharpened.

Screwdrivers must never be used to prise open cases or as cutting tools, though it is in order to keep an old screwdriver for some tasks. Blades can be damaged if a screw is rusted or otherwise jammed in place, so that if reasonable force does not undo a screw, it should be loosened by using one of the many patent lubricants, such as WD40. Sometimes a seized-up screw can be loosed by a hammer blow applied to the screw head with an old screwdriver, but this may cause damage to electronics equipment. By far the better method is to drill out a seized-up screw so that it can be replaced.

Drilling can be done with a hand drill, a portable electric drill or a bench drill. For much servicing work, a battery-operated portable

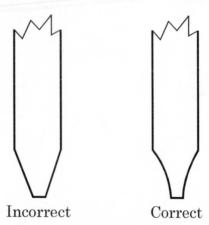

Incorrect   Correct

Figure 2.15   Cross-section of screwdriver.

electric drill can be very useful. The work to be drilled must be securely clamped; never try to hold the drill in one hand and the work in the other. Some electronics boards can be very difficult to clamp adequately, and a better solution is to lay the board on a flat sheet of a material such as Blu-tack. Drills should be matched to the work, and for most electronics purposes a set of carbon steel drill-bits will be suitable. Drills for cutting steel need to be harder, and high-speed drill-bits are more appropriate for such hard metals.

When the drill allows a choice of speeds, use the speed that is appropriate for the type of drill-bit and its size. The larger drill-bits should be used at a slower speed than the smaller ones, and the maker's instructions observed for specialized bit types. Brush away all of the swarf after drilling, and clean up drilled holes with a countersink, a small file, or a wire brush. Sometimes hand use of a larger size bit can be an effective way of cleaning up a drilled hole. Observe safety precautions while using any drill, tying back loose clothing (a tie is particularly dangerous) and using goggles to protect the eyes. Keep both hands well away from the work that is being drilled.

Drill-bits become blunt after use, and should be inspected and resharpened regularly. Resharpening is a skilled job which can be made much easier by using one of the jigs or resharpening tools that can be bought in tool shops. Very great care should be taken to avoid altering the cutting angle of a bit because this can make it quite useless.

Files will be needed in a range of sizes. The larger sizes are unlikely to be used much, but a good selection of miniature files, called 'Swiss files' or watchmaker's files, will be needed, and a few medium cut and final cut files in 6″ and 9″ lengths will be useful. The miniature files will come in for heavier use in servicing work, and need to be maintained well. In particular, a file needs to be cleaned well after being used on softer types of materials, particularly aluminium or its alloys. A wire brush can be used for the larger files, but is less useful for the miniature types.

Never use files as sharp instruments for piercing holes, and never use them as scribers or as levers. The metal of a file is hardened, and a bending force is likely to shatter the metal, a potential hazard to your eyes. When a file becomes too blunt for effective use, replace it because it cannot be resized. Use a wire brush at intervals to clean the 'teeth' of each file, particularly if you have been filing comparatively soft materials, such as aluminium.

Wire cutters, strippers and pliers are often used together, with the pliers used to hold a cable and the cutters cutting it to length. Two or three sizes of pliers should be stocked, with at least one pair of long-nosed pliers. Pliers should not be used as a substitute for spanners, and a set of small spanners, though not listed for the C & G 224 set, will be very useful. Long-nosed pliers should not be used for unscrewing inaccessible nuts, because this is one almost certain way of twisting the jaws of these pliers. The correct solution for such problems is a set of long tubular (box) spanners in the old BA sizes along with some of the smaller metric sizes.

Wire cutters are usually of the popular sidecutter type. These have to be used with care on some types of wire, because the cut fragments will fly from the cutters at high speed and are a hazard to the eyes. The older 'nipper' type of wire cutter is safer in this respect, and one pair should be carried along with the sidecutters.

Wire strippers should be easily adjustable because if the adjustment is clumsy or time consuming there will always be a temptation to try to judge the depth of stripping rather than setting it correctly. Taking a deep bite into a wire in order to strip the insulation will weaken the wire badly, and can be a cause of later failure, particularly on circuitry that is subject to vibration. The insulated wire to be stripped should be held in pliers while the strippers are being used. Both strippers and wire cutters should be replaced if they become blunt or if the blades of cutters become damaged.

Hacksaws used for electronics work can be of the 'junior' size, with perhaps one larger hacksaw reserved for heavier work. A good selection of blades is necessary, because different blades are needed for steel, aluminium and SRBP. The blade must be correctly fitted, with the teeth pointing away from the handle so that the cutting stroke is when the saw is moving away from you. The blade must also be correctly tensioned. A slack blade is likely to twist and snap, and this is a hazard because the snapped blades have very sharp edges.

The work should always be securely clamped, and the saw should be held in both hands, one on the handle and one at the front of the saw frame. Blades should be discarded whenever they are blunt or if they lose a tooth. Blades should be kept in packs so that the different blades can easily be distinguished.

Crimping tools are used to make solderless connections to cables by way of crimped connectors ('crimps') which are compressed on to the wire of a cable. Never use pliers as a substitute for the correct type of tool, because pliers will flatten the crimp rather than close it evenly around the wire. Crimping tools are specialized, and some are not suitable for uninsulated crimps; others are unsuitable for push-on connectors. A good compromise is one combination crimper for both insulated and uninsulated connectors, and a ratchet type for use with insulated crimps.

Marking-out tools are used to mark lines for cutting by hacksaw or for drilling. The scriber is used for marking lines for cutting, and the centre punch for marking a hole position. A centre punch should always be used before drilling so as to ensure that the drill will not run off position. Never use the centre punch as a scriber, however. The scriber should be used in conjunction with a metal ruler for straight lines; for circles a compass scriber can be used. Scribers are sharp, and should be kept in a holder when not in use so that the point is covered. Centre punches should also be kept out of the way. When using the centre punch, the work needs to be laid on a hard surface, preferably a metal block. Use of a centre punch on a soft surface can result in an indentation that is much too large, deforming the metal.

SOLDERING AND UNSOLDERING

Connections to and from electronic circuits can be made by wire-wrapping, crimping, welding or by soldering, and soldering is still the most common connection method for all but specialized equipment. The basis of soldering is to use a metal alloy, usually of tin and lead, that has a low melting point to bond to metals that have a much higher melting point. In a good soldered joint, the solder alloy penetrates the surface of each metal that is being bonded, forming a join that is both mechanically strong and of low electrical resistance.

The action of soldering is straightforward, so much so that it is normally automated in the form of wave-soldering baths, but manual soldering still causes problems. Most of these problems can be traced to inadequate preparation of the surfaces that are to be soldered, particularly when very small-gauge wires are to be soldered. The other main causes of soldering problems concern the under- or overuse of the soldering iron or other source of heat.

To start with, the surfaces that are to be soldered must be clean. The main enemy in this case is oxidation of copper surfaces after storage, and any attempts to solder to a copper surface that is dark brown in colour, instead of the bright gold-red of freshly cleaned copper, will be doomed. For a one-off soldering action, metal surfaces can be cleaned mechanically, using a fine emery paper, but for mass-production some form of acid cleaner or mechanical scrubbing equipment is needed. The tracks of printed circuit boards are particularly vulnerable to oxidation because they have a large surface area. Stranded wire is particularly difficult to clean well, and if a new surface is exposed by cutting a few inches from the end of a stranded cable this will make connection much easier.

In connection with stranded wire, the conventional wisdom at one time was that the end of a stranded cable should always be 'tinned', meaning coated with solder, because it was mechanically connected (such as to a mains plug), so as to avoid the possibility of strands causing a short-circuit. This practice is now frowned on because where some flexing of the wire can happen, as on most connectors, the wire will eventually snap where the tinned section joins the untinned wire. In addition, the clamping screw often works loose as the tinned wire shrinks. A good compromise is to tin only the last millimetre or so of a stranded wire, and insert the wire into a holder so that any fixing screw bears on the untinned wire.

Dirt and oxidation are the main causes of soldering problems but not the only ones. Good soldering requires the solder to be melted on to the whole area of the join, and this is possible only if all of this area is at a high enough temperature. Insufficient heat flow will cause the solder to solidify before it penetrates the other metal, making a join that looks soldered but which is really only a blob of solder placed on the surface. Such a joint is not mechanically strong and is electrically unreliable.

The other extreme occurs when too much heat is used, and the temperature becomes high enough to oxidize the solder and the other metals. Such a joint will often look secure, but it is 'dry', the solder has penetrated but has been boiled out again, leaving a film of oxide with a high resistance. Joints of this sort are a very potent source of trouble because they are difficult to detect and to remedy.

Choice of soldering equipment is important, and it starts with the soldering iron. Despite the name, the tip of the iron is made of copper (sometimes iron plated), and most types of irons come with a selection of different tips for different grades of work. Note that if you use the iron-plated type of bit, it should never be filed to clean it because this will soon remove the iron and allow the remainder of the tip to disintegrate quickly. A lot of work with modern components can be carried out with a miniature electric iron of 12–25 W rating, but there are still a few actions, such as soldering thick bus-bars or metal strips (particularly silver-plated strip lines) that need more heating power than a small iron can provide.

Thermostatically controlled 50 W irons are very useful, but a few soldering jobs are easier with the larger irons that were commonly used in the days of valve circuits. These large irons are rare now, and a useful substitute is the gas-flame torch, using miniature butane cylinders. These need to be used with care, because the gas flame is at a very much higher temperature than the tip of an electric iron. The miniature gas burners are very useful when no source of mains power is available because they are lighter and more compact than battery-powered electric irons. Gas torches can also be used for brazing, which is very much the same as soldering but using a silver–copper–zinc alloy which has a higher melting point. Brazed joints are used where mechanical strength is important, such as chassis joints; its main application on the small scale is to modelling with metal.

The choice of solder is easier. The standard type of solder for electronic uses is 60% tin, 40% lead, in the form of a small-diameter tube that is filled with the flux compound. Flux is a jelly or paste which combines several actions: lowering surface tension to allow solder to spread, protecting surfaces from oxidation, and (to some extent) helping to keep surfaces clean. Typically, the melting point of the solder will be around 190°C, requiring a temperature at the tip of the soldering iron of around 250°C. Virtually all of your soldering requirements will be fulfilled by this type of solder.

For more specialized work aluminium solder and silver solder can be used. Aluminium solder is an alloy that contains about 2% silver in a solder that is much richer in lead than the standard formulation, and this raises the melting temperature to around 270°C, requiring a higher iron temperature or the use of a gas torch. Another option is a solder with 10% silver content which will provide joints of significantly lower resistance. This has a melting point of around 370°C, which is considerably more than most electric irons can provide.

For some small-scale assembly work, solder pots can be used. These are containers that are electrically heated, usually with some provision for scraping the top surface of the molten solder to keep it clean. These pots cannot be used with flux-cored solder, and the usual procedure is to keep the pot topped up with 60/40 solder pellets. The use of a solder pot follows the same pattern as the use of large solder baths in production machinery.

Typically, a soldered join is made as follows. The materials to be joined are thoroughly cleaned, but avoiding the use of corrosive acids that could cause problems later. The joint is secured mechanically if this is possible (two wires can be hooked together, for example) so that the solder is not essential for mechanical strength. The iron is allowed to come up to working temperature, and the tip is cleaned, either by wiping it with a slightly damp cloth or by rubbing it on a proprietary cleaning block such as the Multicore TTC1. The iron is then placed on the joint and after a short pause, the (flux-cored) solder is also placed against the joint. The solder should melt and spread over the joint – it can sometimes be an advantage to move the tip of the iron to help the solder to spread. When the solder forms a thin bright film, remove both the solder and the iron and allow the joint to cool without disturbing it.

Common faults start with trying to work with the iron alone, coating it with solder and then applying it to the joint. This virtually ensures that most joints are made with little or no flux, so that dry and fragile joints result. Another common fault is to dab the iron and the solder on a joint, withdrawing the iron before the solder has spread. This will cause dry and unreliable joints. The opposite, leaving the iron in place until the solder turns dull, will cause severe oxidation and an equally unreliable joint. Continuing to feed solder so that molten blobs of solder drip from the joint is another fault, usually indicating that the metals have not been cleaned thoroughly so that the solder is not spreading.

Desoldering

Very few components on a printed circuit board are contained in sockets, so that components have to be removed by desoldering. A few components can be removed easily, and the simplest are resistors or capacitors whose bodies lie horizontally parallel to the board. It is easy to apply the hot iron to one joint, pull the wire out of the hole in the PCB, and then use the iron on the other lead to pull out the component entirely.

It is not so easy to unsolder vertically mounted components or components with a large number of short leads, such as ICs. For such components you must remove the solder from each joint and then pull out the component (not while you are using the iron). The danger in these desoldering actions is that you will overheat the PCB or the surrounding components, making repairs difficult or impossible.

The two main methods of desoldering are the use of solder braid or the use of a desoldering pump. Solder braid is a form of stranded copper, around 2 mm diameter and usually supplied on a reel. The braid is wound around the joint, and the tip of the hot iron is cleaned, ensuring that it has only a thin file of solder. If a hotter iron temperature can be used than is used for soldering, the process is made easier. The hot iron tip is applied to the braid (not to the solder) and the iron is held in place until the solder of the joint is seen to run into the braid (a form of capillary action). The iron is then removed, followed as fast as possible by taking away the braid. This should leave the joint almost totally free of solder, with only a thin bright film showing. The solder-filled braid is snipped off, and the process is repeated for each joint (with a rest between each to allow the PCB to cool) until all the joints have been treated. The component can then be pulled out.

The desoldering pump is a small cylinder and piston arrangement with a spring-loaded plunger. The plunger is forced in, and the nozzle of the pump held to the joint. The iron is used to melt the solder, and pressing a catch on the pump allows the plunger to return, sucking the solder into the pump. Solder fragments can be removed after several desoldering actions. The pump is easier to use, particularly for close-packed components, and will soon justify its higher cost if a large amount of desoldering is needed. The nozzle of the pump can be either of aluminium or of high-temperature plastic such as Teflon, and nozzles are interchangeable.

Other soldering tools

Heat-shunt tweezers are useful when soldering semiconductors that are susceptible to damage by overheating. They are seldom used in production work, because the flow-soldering methods that are used allow very little heat to flow, and an air blast will cool the components rapidly. For hand soldering and desoldering, thermal tweezers can be clipped on to the wire leads between the body of the semiconductor and the soldered joint. The effect is to allow heat to take the path of least resistance, into the thicker tweezers rather than along the thinner wire, so preventing the temperature inside the semiconductor material from rising to a value which would destroy a junction.

Heat shunts are particularly valuable when desoldering, because the iron often has to be applied for longer, and when components with very short leads are used (as in RF circuits). Components with long leads do not normally need heat shunting.

Soldering toolkits are useful if you do not already have an assortment of tools. Typical sets contain a reamer for cleaning inside plated-through holes and a hook for reaching awkward places. A brush is useful for removing dust and solder fragments, and a fork can be used to hold wires in place. A scraper and a knife are both useful in cleaning wires and PCB tracks. A set of Swiss files is also useful for cleaning work. Avoid the use of acid solutions and old-fashioned flux solutions such as Baker's Fluid.

Wiring up connectors can be tedious, and for all but the simplest and least-used connectors it is worthwhile making up some form of jig that will hold wires in place while they are being soldered. The DIN plugs and sockets are particularly fiddly items, particularly when stranded cables are being used, and it is advisable to tin the ends of the stranded wires to keep the strands together. Preferably, each connection should be clamped into place using small-nosed pliers and then soldered, but some DIN connectors offer no easy way of doing this. Anything you can do to avoid having to juggle simultaneously with a (hot) connector, a cable, solder and a soldering iron must be welcome. A jig is also useful for soldering connections to D-type plugs of the 9-pin or 25-pin type as used for computer serial cables and also for printer leads for PC machines.

Care needs to be taken when soldering cables into coaxial connectors, whether of the TV or the BNC type. It is only too easy to melt the insulation on the cable and to soften the insulation in the connector if the iron is applied too long or in the wrong place. If the connector insulation softens you may find that the inner pin position has changed, and that the connector is unusable. A useful tip is to insert the end of the plug into a spare socket before soldering, to keep the pin in place.

PCB CLEANING

Many fault conditions can be traced to problems on a PCB, and in fact the most common fault found in defective camcorders is damage to a PCB caused by leakage of an electrolytic capacitor. If a leak results in immediate electrical malfunction, the PCB can often be saved by removing the defective electrolytic and cleaning the PCB. If damage has occurred, the same procedure is needed, with the addition of repair to the PCB tracks after cleaning.

The simplest cleaning system makes use of an aerosol spray using 1:1:2 trichloro 1:2:2 trifluoroethylene. This is a CFC type of substance, and should be used with care. The solvent is non-flammable, very volatile, and leaves no deposit. It is also very penetrating, and is by far the most effective cleaning agent for PCBs and also for switch contacts (another common source of trouble). If the use of CFC is unacceptable, an isopropanol spray can be used, but this is flammable and must not be used where there is danger of ignition from sparks or hot objects. A compressed-gas spray aerosol can be very effective for removing dust as distinct from adhering dirt.

When PCB cleaning is required frequently, a better solution is an ultrasonic cleaning bath. This can be used with solvents provided that good ventilation is present, and typical tank sizes are 5 litres and 12 litres. The liquid is vibrated by a transducer operated from a 40 kHz (typical) oscillator, using power levels of several hundred watts, and the effect is to create violent cavitation (alternate compression and evacuation) in the liquid, which carries out a scouring effect on anything placed in the tank. The cavitation also makes evaporation of the liquid much greater, even when the liquid is used cold, hence the need for good ventilation and extraction. Even spaces that are normally inaccessible can be cleaned in this way.

Care needs to be taken that PCBs are suitable for ultrasonic cleaning, since total immersion of a board could remove all the protective wax from components such as capacitors or inductors. The usual system is to place the items in a perforated stainless-steel basket which can be immersed as far as is needed into the vibrating liquid.

REPLACING ICS AND OTHER COMPONENTS

When you are certain that an IC is faulty, because its output is incorrect when all inputs are correct, you will need to remove it and replace it with another identical IC or a pin-compatible type. Your freedom of choice is very limited in this respect as compared to substitution of transistors and other discrete components. An IC which is electrically close enough to use may not have the same connections (or even the same number of pins) as the old one, and in some cases a repair is downright impossible, particularly when a manufacturer has used an ASIC (application-specific IC) that is no longer in production.

A small IC can often be removed by using solder braid on the pins, or a desoldering tool, but in many cases using these methods risks overheating the board or other components. In such cases, and for large ICs, it is often better to snip all the pins close to the body so that the body of the IC can be removed, and then remove the ends of the pins one by one, using pliers and a fine-tipped soldering iron. This has the additional advantage of ensuring that the defective IC is never used again by mistake.

Working on surface-mounting components

Surface-mounting components and boards are now increasingly being used in equipment. The principle of surface mounting is not new – surface-mounting boards for amateur use were on sale in 1977, when they were demonstrated under the name of 'blob-boards' at several exhibitions. The technique, known as SMT (surface mounting technology) has now spread to professional equipment and has resulted in the manufacture of a whole range of components that are designed specifically for this type of fixing. Components for surface mounting use flat tabs in place of wire leads, and because these tabs can be short the inductance of the leads is greatly reduced. The tabs are soldered directly to pads formed on to the board, so that there are always tracks on the component side of the board. Most SMT boards are two sided, so that tracks also exist on the other side of the board.

The use of SMT results in manufacturers being able to offer components that are physically smaller, but with connections that dissipate heat more readily, are mechanically stronger and have lower electrical resistance and lower self-inductance. Some components can be made so small that it is impossible to mark a value or a code number on to them. This presents no problems for automated assembly, since the packet need only be inserted into the correct hopper in the assembly machine, but considerable care needs to be taken when replacing such components, which should be kept in their packing until they are soldered into place. Machine assembly of SMT components is followed by automatic soldering processes, which nowadays usually involve the use of solderpaint (which also retains components in place until they are soldered) and heating by blowing hot nitrogen gas over the board. Solder baths are still used, but the hot-gas method causes less mechanical disturbance and can also allow heat-sensitive components to be shielded.

The construction methods that are used for SM resistors and capacitors makes these components very brittle, and a very common cause of failure is flexing of a PCB. This will often crack the ends of the surface-mounted components, and the insidious problem is that this may not cause immediate failure. Such cracked ends may hang together for years, finally parting after further flexing, perhaps caused by no more than temperature changes. To avoid damage to SM components on a PCB:

- Use a suitable jig to hold the PCB – the PCB should not need to be bent when it is placed in the jig.
- Make sure that the supports for a board do not cause the board

to flex – for example, mounting pillars must all be at the same height.
- Take very great care when inserting or removing daughterboards, as this can cause the motherboard to flex.

Considerable care is needed for handsoldering and unsoldering SMT components, because thermal stress can also damage SM components. A pair of tweezers can be used to grip the component, but it is better to use a holding arm with a miniature clamp, so that both hands can be free. The problem is that the soldering pads and the component itself can be so small that it is difficult to ensure that a component is in the correct place. Desoldering presents equal difficulties – it is difficult to ensure that the correct component is being desoldered, and almost impossible to identify the component after removal; a defective SMT component should be put into a 'rejects' bin immediately after removal.

Chapter 3

Basic Circuits

This chapter illustrates a selection of well-established basic circuits and data, with comments which are reduced to a minimum so as to include the greatest number of useful circuits. The discrete forms of the circuits are shown because these will need to be repaired if faulty; the corresponding IC circuits are dealt with on the usual basis of finding the faulty stage and replacing either the IC or passive components. Circuits of power supplies, including switch-mode supplies and battery supplies, are covered in Chapter 4.

Where several different types of circuits are shown, as for oscillators, practical considerations may dictate the choice of design. For example, a Hartley oscillator uses a tapped coil, but the arrangements for frequency variation may be more convenient than those for a Colpitts oscillator which uses a capacitive tapping. Some crystal oscillator circuits are not always self-starting, particularly with 'difficult' crystals. For this reason, as many variants on basic circuits have been shown as is feasible in the space.

SINGLE-TRANSISTOR CIRCUITS

The basic single-transistor circuits are the common-emitter, common-collector (or emitter-follower) and common-base circuits. The common-emitter circuit of Figure 3.1 is the most widely used form, with the potential divider at the base passing a steady current which must be considerably greater than the average base current – in practice this means from 10 to 100 times the base current. The base voltage is set by this potential divider, and the emitter voltage is developed across the emitter-resistor, whose value sets the collector current for the transistor. If the emitter-resistor is not bypassed by a capacitor, the gain of the stage is set by the ratio of the collector load resistor to the emitter resistor. When the emitter-resistor is bypassed, the gain of the stage approaches the value given by $40V_{bias}$ where V_{bias} is the steady

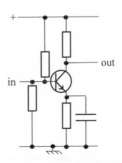

Figure 3.1   Common-emitter single-transistor amplifier stage.

voltage across the load resistor. The circuit provides both voltage and current gain. The input impedance is fairly low, typically a few kΩ, and the output impedance is medium, typically in the region of 10–50 kΩ. The output impedance is effectively in parallel with the collector load resistor, and the input impedance is in parallel with the base bias resistors.

The output waveform is the inverse of the input. The gain and phase values will change considerably for low-frequency signals when the impedance of the emitter bypass capacitor becomes significant in comparison to the resistive value of the emitter-resistor. Gain and phase will also change at high frequencies due to the combined effects of stray capacitances across the load resistor, and the high-frequency cut-off frequency of the transistor itself. Considerable care is needed therefore in substituting types if the transistor is operated at frequencies that approach its cut-off frequency. The cut-off frequency of a transistor operated in this mode is lower than for the other two modes.

The common-collector circuit, Figure 3.2, provides current gain only, with a value of voltage gain that is always less than unity (typically 0.95), with the output in phase with the input. The input impedance is high, with values of several hundred kΩ normally obtained, and higher values can be obtained using more specialized circuits. The output impedance is low, usually less than 100 Ω, and in parallel with the emitter load resistor. Bias can be set by a potential divider or, more commonly, by direct coupling to the collector of the previous stage, Figure 3.2(b).

The circuit is widely used as a buffer, preventing an amplifying stage from being excessively loaded; as an impedance-transformer, and as a current amplifier stage. The frequency response is good, allowing the common-collector circuit to be used at frequencies higher than the cut-off frequency for a common-emitter stage.

The common-base circuit, Figure 3.3, is used mainly in high-frequency amplifier applications or as part of a multi-transistor stage (as in a long-tailed pair or a cascode circuit). The input impedance is very low, a few ohms, and the output impedance is high, so that the effective output impedance of the circuit is usually just the value of the load resistor. The application of high-frequency amplification arises because the high-frequency cut-off for a transistor in this configuration is higher than for common-emitter use. The output is in

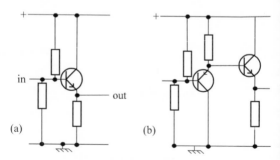

Figure 3.2 Common-collector transistor stage (a) biased by potential divider, (b) biased by direct coupling from previous stage.

48

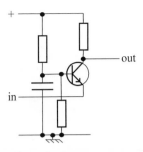

Figure 3.3 Common-base circuit.

phase with the input. Another feature of the circuit is that the collector breakdown voltage for a transistor is highest when the transistor is being used in this configuration.

TWO-TRANSISTOR BASIC CIRCUITS

The obvious methods of connecting two transistors are as a pair of common-emitter, common-collector or common-base circuits respectively. Common-emitter pairs provide a very high gain figure, but the linearity is poor, and such pairs are very often implemented as feedback pairs. The double common-collector pair is known as the Darlington pair, and is extensively used both in discrete and in IC form, but the double common-base circuit is very rare. A further complication of connecting two transistors is the use of complementary NPN–PNP pairs, a method used extensively in IC construction.

Figure 3.4 shows two very common DC connections of common-emitter pairs, shown in their NPN–NPN form without bias details. The shunt-series pair of Figure 3.4(a) provides a gain figure of about

$$\frac{R_4 R_2}{R_3 R_s}$$ where R_s is the source resistance of the driving stage. The

input resistance is low. The voltage gain can be greatly increased by decoupling the emitter-resistor R_3, but this is seldom desirable.

The circuit of Figure 3.4(b) is a series-shunt type, again DC coupled, with a gain of approximately R_4/R_3 if the gain of each transistor is high (i.e. not close to the frequency limit).

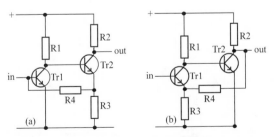

Figure 3.4 Common-emitter pairs (a) shunt-series feedback, (b) series-shunt feedback.

The long-tailed pair, shown in four variants of its bipolar form in Figure 3.5, is the most versatile of all discrete transistor circuits, which is why it is so extensively used in the internal circuitry of linear ICs. The circuit is basically a common-collector stage driving a common-base stage. A *common-mode* signal is a signal applied in the same phase to both bases or gates. Any amplification of such a common-mode signal can only be caused by a lack of balance between the transistors or FETs, so that this value of gain is normally low, often very low. The difference signal is amplified with a considerably greater gain, and the ratio of the differential gain to the common-mode gain is an important feature of this type of circuit, called the common-mode rejection ratio.

The long-tailed pair is most effective when used as a balanced amplifier, with balanced inputs and outputs as in Figure 3.5(a), but single-ended inputs or outputs can be provided for as shown in versions (b) to (d). The overall voltage gain of a long-tailed pair circuit using single-ended input and output is about half the gain that would be obtained from one of the transistors in a common-emitter circuit using the same load and bias conditions. The circuit functions most effectively if the 'tail' portion, represented in the examples by the common-emitter resistor, has a large resistance value, so that versions which use another active component in this position are preferred. Biasing is usually DC, and changes in bias components will cause signal distortion and may also cause overheating. If this type of stage is troublesome, look in particular for changes in the 'tail' resistor (or other current supply), or in the biasing voltages. In general, the IC versions, which are used in virtually all linear ICs, very seldom present reliability problems.

The cascode connection of Figure 3.6 consists of a common-emitter stage driving a common-base stage, using the current gain of

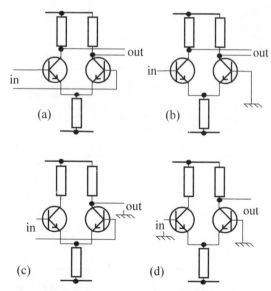

Figure 3.5 The long-tailed pair, in four varieties.

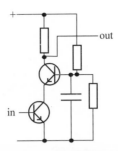

Figure 3.6   Cascode connection of two transistors.

the first transistor and the voltage gain of the second. This provides a high gain-bandwidth figure, and is often used for high voltage outputs because the use of the common-base portion allows the breakdown voltage level to be higher. Cascode stages are therefore used extensively in video amplifiers. The main cause of failure is, as always, changes in value of bias components or breakdown of bypassing capacitors.

The elementary Darlington circuit, Figure 3.7, uses two common-collector stages in sequence, providing a very high input resistance and a very low output resistance. Complementary pairs of this configuration are used more in switching applications than for linear amplification, and the two forms of this type are shown in Figure 3.8. These complementary pairs act as if they were a single transistor of enhanced h_{fe} value, and they have the advantage, compared to the conventional Darlington constructed from identical transistor types, of having a lower DC voltage between input base and the effective emitter connection, since there is only one junction rather than the two for the conventional Darlington.

AUDIO CIRCUITS

Audio equipment of more recent construction will consist almost exclusively of ICs, but older equipment, and certainly equipment that can be classed as Hi-Fi (in the best sense of that abused description) will feature discrete circuits which are often ingenious and sometimes dependent on the peculiar characteristics of one type of transistor.

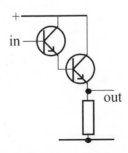

Figure 3.7   A typical Darlington circuit.

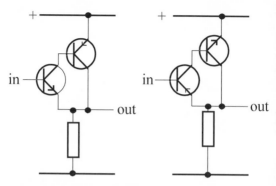

Figure 3.8 Complementary Darlington circuits.

Finding and curing faults on such equipment can often be difficult unless good service sheets and suitable replacement devices are available. Note that the following is concerned with basic circuits only, audio circuits of more specialized types as used in Hi-Fi equipment are covered in Chapter 6.

Figure 3.9 illustrates a magnetic pickup preamplifier circuit for a vinyl disc player. This is a circuit that was extensively used in the latter days of vinyl discs, and which is likely to be found in equipment that is now requiring service. The important feature here is the frequency correction or equalization circuitry that is needed for disc replay. Discs are recorded to the RIAA standard (see BS 1928:1965) in which bass frequencies are attenuated to prevent excessive cutter movement and treble frequencies are boosted to ensure that their recorded amplitude is well above the level of surface noise. This deliberate distortion must

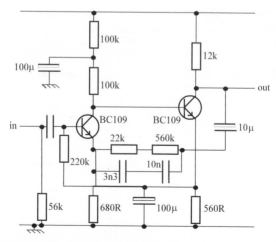

Figure 3.9 Magnetic pickup preamp design.

be corrected at playback by using *CR* networks or their equivalent; to achieve this correctly three time constants of 75 μs, 318 μs and 3180 μs are used. The 75 μs time constant is that of a treble-cut filter taking effect with a turnover point of 2.12 kHz; the 318 μs time constant is a bass boost with a turnover of 500 Hz, and the 3180 μs is a final stage of bass cut at 50 Hz and under to reduce rumble noise.

The output from most magnetic cartridges other than the moving coil type is around 5 mV at the standard conditions of 5 cm/s stylus velocity and 1 kHz signal. This corresponds to a signal amplitude of close to the minimum 2.5 mV that most amplifiers need to give full output at full volume-control setting. The preamplifier should present an input resistance of around 50k and should be capable of accepting considerable overloads at the input, 50 mV or more, without noticeable distortion. The circuit in this example is a series-shunt feedback pair with the equalizing carried out in a separate feedback loop, and faults in this stage are most likely to be caused by resistance changes which affect the bias levels, or by bypass capacitors becoming leaky.

The trend, before vinyl disc players started to be phased out in favour of CD units, was away from the feedback loop type of equalization and towards an equalizing network of purely passive components applied following a 'flat' preamplifier stage – this is held to be advantageous because of the effects of transients on feedback amplifiers, particularly when momentary overloading is caused. Such circuits are rather easier to work on.

Figure 3.10 shows some typical basic tape/cassette circuits. Once again, equalization is needed, but not all the time constants are so rigidly fixed; they tend to change as new tape materials and new types of tape-head construction emerge. In addition to the standard corrections, individual tape decks may need further corrections, a multiplex filter may be included to remove FM stereo subcarrier signals, and noise-reduction circuits such as Dolby or dbx may be used, virtually always in IC form. At the last count, standard equalization frequencies that were being used on replay were 3180 μs for all tapes, and either 70 μs or 120 μs for chrome and ferric tapes respectively, with ferrichrome and pure iron tapes replayed at 70μs. The equalization that is needed for recording amplifiers is more specialized because recording equalization time constants depend much more on individual needs determined by the tape-head construction.

Figure 3.11 shows a typical preamplifier stage for a moving coil microphone. The particular features here are low noise operation matching a fairly low input resistance, high gain, and hum rejection. The low output and low source resistance of the moving coil

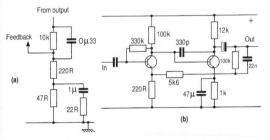

Figure 3.10 Typical tape/cassette inputs (a) feedback equalization components, (b) cassette recorder input stage.

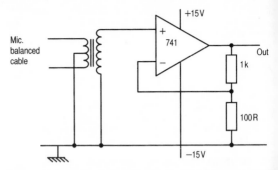

Figure 3.11 Typical moving coil microphone input preamplifier stage.

microphone requires the use of a microphone transformer (or a built-in FET amplifier). If a balanced layout is possible, hum pickup can be greatly reduced. This circuit is likely to be found only in older equipment, and IC input stages are used for applications such as computer sound cards.

Figure 3.12 shows one version of a circuit which has become virtually the standard method of tone control used in audio systems, the Baxandall circuit which uses a feedback loop to contain the control components. There is very little interaction between the treble and the bass controls, low distortion, and a good range of control amounting to 20 dB of boost or cut. The Baxandall circuit is usually placed between bipolar transistors but several designs claim significant improvements by using a FET at the output stage, and the circuit is also used with IC stages.

Tone controls and equalizing circuits, along with other filter requirements, can make use of active filters, of which some examples of low-pass, high-pass, bandpass and notch filters are illustrated in Figure 3.13. These designs use only resistors and capacitors for the filtering action together with semiconductors to provide gain for a feedback loop, and they are considerably easier to design than *LC* filters, with more predictable characteristics. The filters generally have a slope of 12 dB per octave, meaning that the response changes by 12 dB for each doubling or halving of frequency.

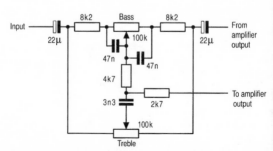

Figure 3.12 The well-known Baxandall circuit.

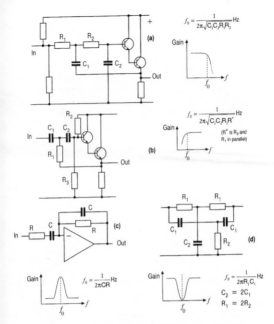

Figure 3.13 Some active filter circuits.

AUDIO OUTPUTS

Class A stages are those in which the transistor(s) is/are always biased on and never saturated (bottomed). A Class A stage can be used for better linearity, but the poor efficiency (less than 50%) of a Class A stage means that the heat dissipation is high, and great care needs to be taken over cooling. Failures of Class A stages are most commonly due to overheating, and it is never enough to repair the damage – you need to let the customer know that it's not a good idea to have the amplifier placed against a radiator. One point to remember is that the thermal dissipation of a Class A stage is independent of signal content, so that a Class A amplifier is less likely to be overheated by excessive output than a Class B stage. Owners of Class A amplifiers are, in any case, less likely to play the sort of stuff that causes overloads.

The simplest Class A stage can use a transistor with an output transformer, but such circuits, though used in some early all-transistor TV receivers, are rare now, and the more usual configuration is to use two or more transistors which share the current in some way (a *push-pull* stage), with no transformer. The coupling to the loudspeaker will either be through a large-value electrolytic (a component that should be checked in the event of problems in this type of stage) or using a direct coupling circuit with elaborate DC feedback and protection networks to prevent DC flowing in the loudspeaker coils. Failure of these latter circuits can be difficult to pinpoint because of the inter-actions of the feedback loops, and all checking and testing should be done using a resistive dummy load rather than the loudspeaker.

The heatsinks for Class A circuits will be of large area, and should be clean – a dirty heatsink can make the difference between normal operation and thermal overloading. If an output transistor has to be replaced, make certain that you follow the manufacturers' recommendations on the use of heatsink grease to ensure that the thermal resistance is as low as possible.

Class B audio operation uses two (or more) transistors biased so that one conducts on one half of the waveform and the other on the remaining half. Some bias must be applied to avoid crossover distortion which is caused by the range of base-emitter voltage for which neither transistor would conduct in the absence of bias. Class B audio stages can have efficiency figures as high as 75%, though at the expense of rather higher distortion than a Class A stage using the same layout. The higher efficiency enables greater output power to be obtained with smaller heatsinks, and the use of negative feedback can, with careful design, reduce distortion to negligible levels. The dissipation depends on signal content, so that overloading can cause excessive heat dissipation, though modern circuits provide for elaborate protection circuits which themselves can be a fruitful source of trouble when components fail.

Figure 3.14 shows the 'totem-pole' or single-ended push-pull circuit, which can be used for either Class A or Class B operation according to the bias level. This version uses the more common Class B circuit using complementary symmetry – the output transistors are PNP and NPN types. This has been a circuit with a very long life, and has been in use for output stages for 30 years or so. Hi-Fi purists have noted that the circuit is not ideal because of the differences between PNP and NPN power transistors of apparently compatible characteristics, and some of the variations on this theme are noted in Chapter 6. The main alternative, the pseudo-complementary circuit, such as that of Figure 3.15, uses the PNP-NPN pair working as drivers, and allows the use of a matched pair of NPN transistors for the output stage. This also is not ideal for the highest levels of Hi-Fi.

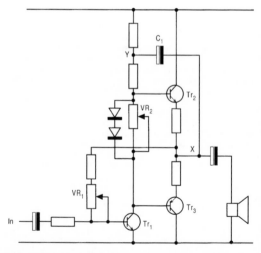

Figure 3.14 Typical single-ended push-pull circuit.

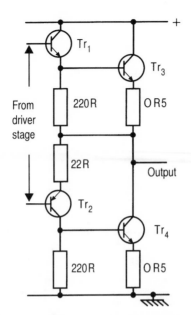

Figure 3.15 The pseudo-complementary circuit.

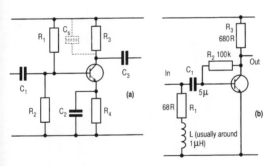

Figure 3.16 Elementary frequency compensation circuits.

OTHER CIRCUITS

Wideband amplifiers can range from the video amplifier for monochrome TV to the specialized circuits used in oscilloscopes, and the methods that are used to extend bandwidth must include the use of transistors whose gain-bandwidth figure is high – replacement of a transistor in such a circuit must be made with care over selection of an identical or compatible type. The circuit methods that are used include 'peaking' (frequency compensation), and negative feedback to trade gain for bandwidth.

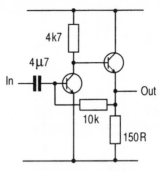

Figure 3.17 A typical feedback pair circuit used for video frequencies.

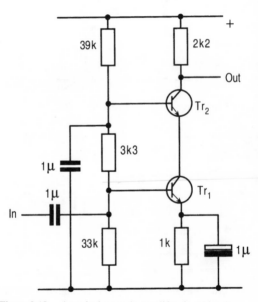

Figure 3.18 A typical cascode amplifier for video frequencies.

Figure 3.16 illustrates two basic methods of frequency compensation using inductors or capacitors to compensate for the shunting effect of stray capacitances. In Figure 3.16(a), the capacitor in the emitter circuit forms a time constant with the emitter-resistor which equals the time constant of the collector load with stray capacitances. Figure 3.16(b) shows the use of an inductor-resistor circuit in the base input, using the inductor to resonate with the input capacitance. Another device, not illustrated, is to use an inductor in the collector circuit. Of

58

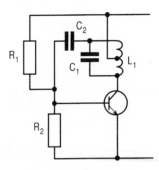

Figure 3.19 The Hartley oscillator circuit.

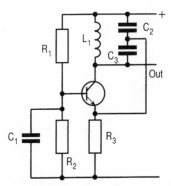

Figure 3.20 The Colpitts oscillator circuit.

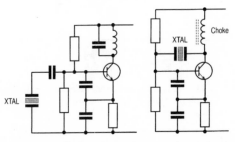

Figure 3.21 Typical crystal-controlled oscillator circuits.

these, the circuits using inductors have fallen out of favour except in TV circuits because of the problems of 'ringing' caused by resonances with other capacitances, and the trend away from the use of inductors has now made their appearance, even in TV receivers, a rarity.

Figure 3.17 shows a shunt-series DC feedback pair circuit, one of

59

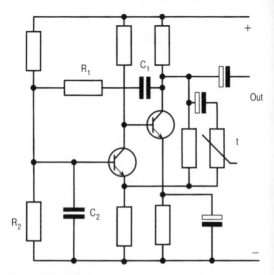

Figure 3.22 The basic Wien bridge circuit, used mainly for low-frequency oscillators.

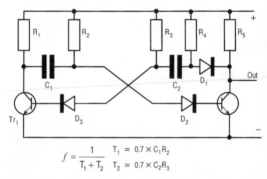

$$f = \frac{1}{T_1 + T_2} \qquad T_1 = 0.7 \times C_1 R_2$$
$$T_2 = 0.7 \times C_2 R_3$$

Figure 3.23 A typical multivibrator circuit using transistors, showing the diode clipper at the output which improves the waveshape.

the basic transistor pair circuits, which is a useful format for video frequencies, often used with some frequency compensation. Figure 3.18 shows a cascode amplifier which has the advantage of high stability, because there is practically no feedback from output to input and high gain over a large bandwidth. This circuit is extensively used for video output stages because it also has the advantage of allowing the use of high collector voltages. FET cascodes and combinations of FET and bipolar transistors can also be used.

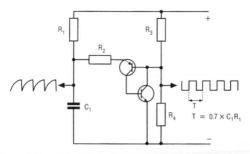

Figure 3.24 The serial multivibrator circuit.

$$T = 0.7 \times C_1 R_1$$

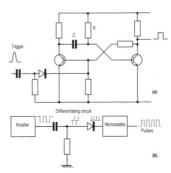

Figure 3.25 A monostable in transistor form (a) and the block diagram (b) of a pulse generator using an astable and a monostable.

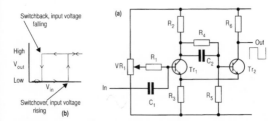

Figure 3.26 The Schmitt trigger and typical input/output graph showing hysteresis.

Oscillators for computing circuits need not be sine-wave types, but for radio and TV use sine-wave oscillators operating at high frequencies are needed, and can be provided in IC or discrete form. The Hartley type of oscillator (Figure 3.19) uses a tapped coil; the Colpitts type (Figure 3.20) uses a capacitor tap.

Although these are not the only r.f. oscillator circuits, they are the circuits most commonly used for variable frequency oscillators. Figure 3.21 illustrates two versions of the Colpitts crystal-controlled oscillator circuit. The frequency of the output need not be the fundamental crystal frequency, since most crystals will oscillate at higher harmonics (overtones) and harmonics can be selected at the output. Frequency multiplier stages can then be used to obtain still higher frequencies. These circuits have now mainly been replaced by frequency synthesizer stages using ICs.

For low frequencies, oscillators such as the Wien bridge (Figure 3.22) or twin-T types are extensively used. The useable frequency ranges are from 1 Hz, or lower, to around 1 MHz, and such circuits are widely used in older signal generators, particularly types used in school electronics courses. The amplitude of signal has to be stabilized by using a thermistor in a feedback loop, and failure of this thermistor or other components in the loop will cause severe distortion of the waveform.

Untuned or aperiodic oscillators are important as generators of square and pulse waveforms. Figure 3.23 shows the familiar multivibrator (astable) together with the modifications that are needed which improve the shape of the waveform. The less familiar serial multivibrator is shown in Figure 3.24; this circuit is a useful source of narrow pulses. Although an IC generator will be used in any modern pulse generating equipment, older (particularly school) equipment is likely to use the discrete forms. With so few components, failure is usually easy to diagnose and repair.

When a pulse of a determined, or variable, width is required from any input (trigger) pulse, a monostable circuit must be used. Figure 3.25 shows a monostable circuit, with a block diagram to illustrate how the combination of astable and monostable forms a variable-frequency pulse generator. As usual, these circuits nowadays use ICs, but they can be found in discrete form in older equipment. If there is ever a demand for servicing vintage radar equipment, the valve versions of these circuits will be easily recognizable.

The basic bistable circuit is most unlikely to appear in discrete forms, and this applies also to the Schmitt trigger which gives a sharply changing output from a slowly changing input. The hysteresis (voltage difference between the switching points) is a particularly valuable feature of this circuit, illustrated in Figure 3.26. A circuit with hysteresis will switch positively in each direction with no tendency to 'flutter' or oscillate, so that Schmitt trigger circuits are used extensively where electronic sensors have replaced purely mechanical devices such as thermostats. These have, in turn, been replaced by microprocessor-based circuitry in which an output from a thermistor is sampled at intervals and used to alter the heating conditions only when several readings are mostly either below or above the set value.

PROBLEMS OF NOISE AND INTERFERENCE

The noise level of a resistor is specified in terms of microvolts (μV) of noise signal generated per volt of DC across the resistor. Such noise levels range from 0.1 μV/V for metal oxide to a minimum of 2.0 μV/V for composition, and for composition resistors the value increases for higher values of resistance. Linear semiconductor stages are a more potent source of noise, but the most critical components in any amplifier are the resistors at the input of the first semiconductor stage, because their noise will be amplified by all the following stages. For this reason, resistors in the first stages of any amplifier must, if faulty, be replaced by low-noise types as specified by the manufacturer of the equipment.

When equipment comes for repair with a complaint of noise, the search for the source should therefore concentrate on the early stages unless there are case examples in the service sheets to indicate other causes. Apart from the resistors at the input, excessive bias current in an early semiconductor stage can also be a potent source of noise, along with poor connections at the input. A poor TV signal input can often be traced to a loose-fitting coaxial plug, and it can surely only be a matter of time before we adopt the types of screw-fitting coaxial connectors that are used in US cable adapters and receivers.

Complaints of noise in equipment need to be checked, because the consumer cannot necessarily be expected to know where the noise originates. It is not unusual to find a TV receiver being returned on the grounds of a high noise level ('the screen is all whirly', was one description) only to find that the user watches videos and that the video heads need cleaning. In addition, the noise may well originate from a defective aerial or distribution amplifier, so that the equipment should be roughly checked in place before being taken to the workshop.

Noise is a much less common complaint in audio equipment, and in any case is often inaudible compared to the sound output, so that noise problems are likely to arise only from customers who play soft music and have delicate eardrums. The source of noise will usually be located in input stages, and it is very unusual to find noise present on all inputs – the usual problems centre on moving coil pickup inputs for vinyl discs. CD inputs are less likely to cause noise problems because the conversion from digital to analogue format takes place at a comparatively high signal level.

Interference can be natural or man-made. Many of the complaints of TV interference arise from the natural sunspot conditions that allow co-channel interference, and there is nothing that can be done. TV weather reports usually warn of such conditions, but not all viewers see the reports. The other main natural cause of interference is thunderstorms, and most people are by now accustomed to the sort of interference that this produces, and are unlikely to call for servicing as a result of this type of interference.

Man-made interference is likely to be more avoidable. The most widespread example is likely to be on the startup of Channel 5, though the programme of retuning video recorders should keep complaints to a minimum. Nevertheless, there will inevitably be a number of viewers who slip through the net and who will make service calls when the new programme starts.

The design of modern car ignition systems has made this potential source of interference a negligible cause of complaints nowadays, but the use of older vehicles, encouraged by the abolition of road tax on anything more than 20 years old, may bring a few examples of unsuppressed engines on to the road, causing a trail of interference complaints. This should always be referred to the authorities (see later) because nothing can be done by alterations to the receiver.

The usual mechanism for interference is the presence of a strong signal which will beat with the wanted signal to give an IF resultant. For example, if you are using an FM radio tuned to 100 MHz, then a transmission at 110.7 MHz will give an output at 10.7 MHz which can be strong enough to break through the input-tuned circuits to the IF stages. There are no legal transmissions which can do this, and none at 10.7 MHz, so that such interference is virtually always caused by illegal transmissions, and if the signal strength is enough to cause interference, the source must be close enough. Such problems should be reported.

Breakthrough interference from transmitters is always a problem, particularly from taxi two-way radio. Providing that the transmitters are on frequency and being used at the legal power levels, nothing can be done other than to try to improve signal trapping at the affected receivers, and this can be a tedious cut and try process unless ready-made filters are available.

Interference to FM radio from other transmissions is usually confined to car radios unless a receiver is being used in a region of very poor signal strength. It is too much to expect a cheap radio receiver operated from a short rod aerial to be immune from interference, and customers in general appreciate this. When interference affects a Hi-Fi installation which is connected to a good aerial system, then this is a case for reporting the problem rather than trying to tackle it for yourself.

At one time, all complaints of radio interference could be investigated by the Post Office, but the situation is now much more complex. The Department of Trade and Industry (DTI) has a Radio Investigation Service (RIS) which covers the use of illegal transmitters or licensed transmitters being used in an illegal way (at higher power output or incorrect frequency). A fee, currently £21, will be charged for any call on the RIS, and no action will be taken unless you can produce a log that details the time, duration and type of interference experienced. No action will be taken unless your equipment is using a good outdoor aerial. This latter step is essential because of the number of complaints that arise from listeners or viewers who have bought what are described as 'miraculous' indoor aerials and found, too late, just how remarkably useless they are. Persistent interference from a car ignition system should be reported to the police.

Interference from licensed and correctly operated transmitters is not the concern of the RIS, and should be referred to the manufacturer of the receiver or his local agent. The thinking behind this is that receivers that are well designed with filtering at the input, and connected to a good aerial system, should not experience interference from a correctly operated transmitter. Such interference can usually be attributed to poor maintenance, incorrect tuning or a low-gain aerial, none of which is the concern of the RIS.

CHAPTER 4

POWER SUPPLIES

SIMPLE MAINS SUPPLIES

A simple mains power supply (as distinct from switching types, see later) requires the mains AC to be transformed to the correct voltage level, rectified to unidirectional current, and then smoothed so as to supply DC. The type of equipment that is required depends very much on the levels of voltage and current that are required and in this book we are mainly concerned with the circuits designed for the conventional levels of current and voltage that are involved in consumer electronics equipment, that is, the levels required for IC and transistor circuits.

- Power supplies were at one time comparatively easy to work on and it was easy to deduce the cause of failure. Modern circuits are by no means so simple, and in this chapter emphasis has been placed on basic circuits and the way that elaborate ICs operate so as to help in locating faults.

At low levels, meaning voltages of up to 50 V and currents of up to a few amps, the standard methods make use of silicon junction diodes in bridge form, and electrolytic capacitors, with low-voltage AC being provided by way of small transformers. Higher current supplies demand diodes that are mounted on heatsinks, along with the use of Schottky diodes which have lower forward voltage drops, and higher voltage levels are catered for by silicon diodes up to considerable current levels. Specialized EHT silicon diodes can be used for voltages as high as 7 kV RMS per diode.

The essential features of a power supply are:

- The off-load and on-load voltage levels.
- The maximum continuous load current.
- The output resistance which is equal to:

$$\frac{\text{voltage change between no-load and full-load}}{\text{maximum rated on-load current}}$$

- The regulation, which measures the effect of mains voltage variations, is:

$$\frac{\text{no lead volts} - \text{full load volts}}{\text{full load volts}} \%$$

Rectifiers

The most common form of rectifier device is the silicon diode. The threshold voltage, below which no current flows in the forward direction, is about 0.6 V and is a quantity that is determined by the nature of silicon and the use of a PN junction rather than by the construction. Once conducting, the shape of the V/I characteristic is exponential so that a considerable increase in current is required to increase the voltage drop across the diode, as if the diode had a variable amount of internal resistance whose value decreases as the current increases. Manufacturers will usually specify the value of forward voltage at various values of forward current, or quote an 'average' value of slope resistance.

In the reverse direction, the silicon diode is non-conducting with negligible leakage current until the reverse breakdown voltage is reached. At the reverse breakdown voltage, the amount of current that can flow is controlled only by the total resistance in the circuit, so that reverse breakdown is usually destructive, causing enough current to flow to overheat the diode and burn it out. Because of the importance of avoiding reverse breakdown, diodes are usually rated for V_{RRM}, the maximum reverse voltage that will be applied. This is not the same as the RMS voltage of the AC that is being rectified, and its value depends on the type of rectifying circuit that is used, see later in this chapter.

Because of the forward voltage, a silicon diode dissipates power, and the amount of power in watts is given by:

$$P = V_f \times I_f$$

where V_f is in volts (usually around 0.8 V) and I_f is in amperes. For example, if the forward voltage at 5 A is 0.8 V, then the power dissipated is $0.8 \times 5 = 4.0$ W. The diode must be capable of dissipating this amount of power, otherwise overheating will occur and a heatsink will have to be used, requiring the silicon diode to be of the stud-mounted variety. This heatsink should be kept clean to avoid overheating.

In addition, care must be taken that the peak ratings are not exceeded. A diode may be used at a current which on average is within the rated value but which consists of current peaks which might be on the limit of the ability of the diode to withstand. Most silicon diodes can withstand very large current peaks provided that these are brief, but where very low transformer winding resistance values exist and the diode is connected directly to a low-resistance electrolytic (see later for capacitor ESR) then it is possible to exceed the peak current rating of a diode, causing breakdown and failure. Do not assume that an unobtainable silicon diode can be replaced by another of the same reverse-voltage and forward-current ratings.

The normal junction silicon diode is often replaced for high-current rectification applications (as in personal computers) by Schottky diodes using N-type silicon in a junction with metallic aluminium. Unlike the PN junction, a Schottky junction uses only one polarity of carrier, electrons in this case. A more important effect is that the forward voltage is significantly lower even at high currents. Typical examples from the Motorola range include diodes with forward voltage levels of 0.475 V at 3 A, up to 0.78 V at 300 A for a matched pair of high-current diodes. These should be compared with voltage levels of 1.4–1.6 V for silicon junction diodes of the same current capabilities. Schottky diodes that have failed must not be replaced by silicon diodes.

Old equipment, particularly older radio and TV receivers, may still contain selenium rectifiers. These should *always* be replaced by silicon diodes in the course of servicing, if the equipment is likely to remain in use, and the selenium rectifiers should be sent to a specialist for recovery of the selenium. These rectifiers provided a long and useful life, but when they failed and selenium was burned, the fumes were dangerous and for safety reasons they should be replaced on any equipment that is likely to be used unless historic interests require all of the circuit to be as it originally was.

BASICS

For single-phase circuits, silicon diodes are used in half-wave, bi-phase half-wave, and full-wave bridge circuits, (Figure 4.1), of which the full-wave bridge is the circuit most likely to be used other than for low-current applications. The output of each rectifier circuit is a set of half-waves of roughly sine shape and requires smoothing. Table 4.1 shows the output as measured by a DC meter for smoothed outputs, given an input of E volts peak ($0.7E$ volts RMS). Failure of one diode in a bridge, or failure of a smoothing capacitor, is marked by a notable reduction in output voltage to the Max.-load value shown in the table.

The waveform from a full-wave bridge or bi-phase half-wave rectifier circuit is, if diode drops are neglected, a set of half-cycles in one direction. This already has a considerable DC content, but also has an unacceptable high ripple (alternating content) for most electronics applications. For some thyristor circuits, however, a raw

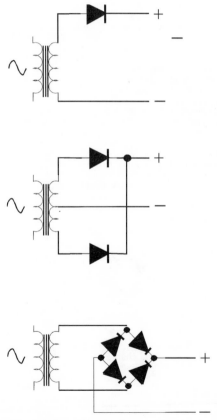

Figure 4.1 Diode rectifier circuits.

Table 4.1 Outputs from rectifier circuits as measured by a DC meter

Circuit	No-load output	Reverse volts	I_{dc}/I_{ac}	Max.-load output
Half-wave	E	$2E$	0.43	$0.32E$
Bi-phase	E	E	0.87	$0.64E$
Bridge	E	E	0.61	$0.64E$

rectified supply is essential in order to allow the thyristor to attain a non-conducting state.

The most common method of smoothing the output from such a rectifier is by the use of a large reservoir capacitor. The name aptly describes the action. The reservoir capacitor is charged as the diodes conduct, reaching the peak DC value (less diode voltage drops). As the output waveform drops to zero again, the diodes will cut off, leaving the reservoir capacitor storing the peak voltage value. If there is no leakage current, the capacitor will remain storing this voltage, and the diodes will not conduct again – in practice, with no load, the diodes will conduct sufficiently to keep the capacitor charged to the peak voltage of the AC output from the transformer, (Figure 4.2).

The addition of this capacitor has two effects. One is that the diodes conduct briefly at around the peak voltage, so that the diode current is flowing for very much less time than the half-cycle that is used in the unsmoothed circuit, and so needs to be greater in value than the average current. The other effect is that the reverse voltage across each diode is much greater if the half-wave or bi-phase half-wave circuit is used. In the full-wave bridge circuit, this reverse voltage affects two diodes in series, and if the reverse leakage currents are matched, each diode will experience only the AC peak voltage. It would, however, be rather foolhardy to assume that the reverse voltage always affects the two diodes exactly equally.

When a load is connected, the situation becomes considerably more complex. Unless the load current is almost negligible, the addition of a load will cause the output DC voltage level to drop, and the diodes will conduct between this level and the peak value, passing current

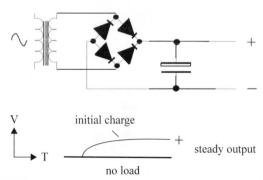

Figure 4.2 The effect of a reservoir capacitor with no load.

that both supplies the load and recharges the reservoir capacitor. For all of the time when the AC output level is below the DC output level and the diodes are cut off, the reservoir capacitor *by itself* supplies the current, so that the voltage across the reservoir capacitor will drop exponentially, as illustrated in Figure 4.3. The rise and fall of voltage follows part of a sine-wave shape during charging, and part of an exponential discharge at other times, so that its appearance is approximately that of a sawtooth. This wave constitutes the ripple voltage, and most texts assume that it is a sine wave whose RMS value can be given and whose frequency is twice mains frequency for full-wave or bi-phase half-wave.

These assumptions are not correct, and precise measurements of ripple are by no means simple. In addition, the shape of the ripple, though its fundamental frequency is at twice mains frequency for full-wave or bi-phase half-wave (mains frequency for a half-wave circuit), contains harmonics which if not suppressed can cause a considerable amount of interference. These harmonics are usually dealt with by other smoothing components, but their presence should be kept in mind; they cannot simply be neglected. Small chokes used in power supplies are intended to suppress higher ripple frequencies, and should not be removed.

Drawing current from a simple rectifier and reservoir capacitor supply will therefore cause a reduction in the DC voltage output and an increase in the ripple amplitude, both caused by the reservoir action alone. In addition, the diode drop and the effect of series resistance in the circuit will cause a further drop in the DC voltage level as load current is supplied. The no-load output voltage will be equal to the AC peak voltage less diode drops, and the on-load voltage will lie somewhere between the peak value and the value for an unsmoothed supply, which is 0.885 × RMS AC value, or 0.632 × peak value for a full-wave rectifier. If we take it that the worst unsmoothed output is 60% of smoothed output for a full-wave bridge of bi-phase half-wave rectifier, then the output value at any practical load value must lie within these limits, ignoring diode drops.

No really complete analysis is sufficiently easy to use to be of practical interest, but Figure 4.4 shows a graph in which the output voltage is plotted against the time constant of the reservoir capacitor and the load resistance. This, though produced purely by approximations, can be a useful guide to output voltage and its variation. The

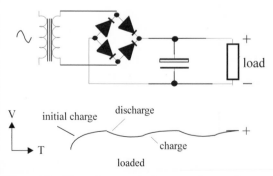

Figure 4.3   The voltage across the reservoir capacitor on load.

output voltage is plotted as a percentage of the peak no-load voltage, and the smoothing constant as the fraction CR/t, where C is reservoir capacitance in farads, R is the load resistance in ohms, and t is the time between peaks in seconds – this will usually be 0.01 s (10 ms) for a 50 Hz supply and full-wave rectification. For a 60 Hz supply and full-wave, the time is 0.0083 seconds. Diode drops are ignored.

The other factor that needs to be considered in a power supply of this type is the current rating of the diode. The recharging of the capacitor takes place over a comparatively short time in the cycle, considerably shorter than the time for which the capacitor supplies the load. The amount of current that is drawn in order to charge the capacitor as well as to supply the load for this short time is therefore considerably greater than the amount of average steady current. Fortunately, silicon diodes have large peak-current ratings, and the pulse nature of the current is seldom a problem. The peak-current rating of diodes is not always quoted by manufacturers, but is usually more than 10 times the average current rating for a 10 ms time. As an example, RS Components quote for their stud-mounted rectifier 16 A diodes a peak surge of 230 A, but this is for a switch-on surge of 10 ms, not for repetitive surges.

The internal resistance of a power supply of this type is not a constant, and is not equal simply to the measurable resistances of the transformer windings and the diode. The drop of voltage as load current is taken from the reservoir is the equivalent of an internal resistance, but its value is not constant and not calculable to any degree of accuracy.

VOLTAGE MULTIPLIERS

A voltage multiplier circuit provides a DC output whose level is higher than the peak AC voltage from the transformer or other AC source.

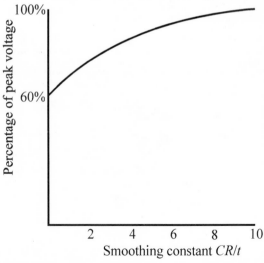

Figure 4.4 A rough guide to the effectiveness of the reservoir capacitor.

The simplest form of circuit, the half-wave doubler, is illustrated in Figure 4.5. If we imagine this at the point of being switched on with all capacitors uncharged and no load, then the first positive half-cycle will cause D2 to conduct, charging C2 to the peak of the AC voltage. When the AC voltage reverses, D1 will conduct, and this will now charge C1 to the peak voltage, positive at the cathode of D1. The next positive cycle of AC will now be superimposed on to the existing voltage at the cathode of D1, equal to peak voltage, so that on the next positive cycle, C2 will be charged to twice the peak voltage of the AC. This capacitor must therefore be rated for at least twice AC peak voltage working. From now on, the voltages that are maintained will be equal to peak AC at the cathode of D1 and twice peak AC at the cathode of D2.

The output resistance of a voltage doubler is usually high, mainly because the capacitors have a fairly large reactance even at the higher supply frequencies which are almost always used along with multiplier circuits. Unlike the rectifier circuits for low-voltage supplies, there is no real lower limit to the output of a multiplier circuit on heavy load and if the load resistance fluctuates considerably some stabilization will be needed. Most multiplier circuits make use of waveforms derived from oscillators, and a control circuit for stabilization samples the output voltage and uses this to control the amplitude of the oscillator. As for any other rectifier-reservoir type of supply, the ripple amplitude increases as the DC output level drops.

Figure 4.6 shows an alternative form of doubler, in which one lead from the transformer is connected to the junction of two capacitors. This configuration has the advantage that each capacitor is subjected to only the AC peak voltage unlike the second capacitor in the circuit of Figure 4.5. The AC input in this design is floating with neither point earthed, which can make insulation requirements easier, and it ensures that the diodes also share the peak-reverse voltage.

Voltage tripler and quadrupler circuits, (Figure 4.7), can make use of the same principles of using a capacitor to charge by using current from one diode and then to superimpose this voltage on to another conducting diode along with an AC half-wave. The more the number of stages of multiplication, the poorer the regulation, but for load currents which in many cases do not exceed 1 mA this is usually unimportant. For TV receiver use, voltage multipliers make use of the pulse waveforms in the line-output transformer, and the use of miniature EHT silicon diodes allows a combination of winding and rectifier to be used, avoiding the need for high-voltage capacitors. The capacitance that is used in the circuit is in fact the stray capacitance of

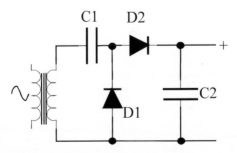

Figure 4.5   The simple voltage-doubler circuit.

71

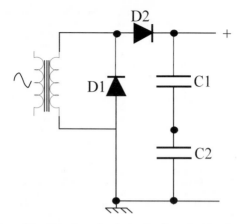

Figure 4.6 The alternative form of voltage-doubler circuit.

the carbon coating on the outside of the picture tube. Poor connection to this coating can result in a low EHT voltage.

STABILIZATION PRINCIPLES

Stabilization means the maintenance of a fixed level of voltage or (less commonly) current from a power supply. Though stabilization is usually nowadays associated with IC stabilizers, these provide just one method out of many, and some other methods have peculiar advantages for special purposes. The servicing of stabilized power supplies is not totally confined to replacing IC stabilizer chips, and when a stabilized supply starts to give trouble, it is sensible to check the electrolytic reservoir capacitor before worrying about the IC.

A stabilizer operates by wasting power, and it is usually needed to stabilize against two causes of variation of output. The first is alteration of the mains supply (the regulation effect), caused by sudden loads (lifts, electric kettles being switched on after the end of '*Coronation Street*', etc.). A good stabilizer should not pass any of these voltage surges to the DC output, and in some cases, particularly impulse interference, more than simply stabilization is needed. The other effect is the variation of voltage caused by the variable load current passing through the internal resistance of the power supply, including the effect of the reservoir capacitor. A stabilizer should be able to keep the output terminal voltage steady despite current variations, slow or fast, between full-rated load and zero load.

The peak voltage output from a power supply will fall to some lower value at full-rated output current, and this lower value is the one which can be stabilized. In fact, to allow for the effect of the stabilizer itself, the output voltage will be lower still. This implies that the amount of waste can be considerable. For example, a 5 V DC output from a stabilizer will require an input to the stabilizer of about 6 V *absolute minimum*, and in order to provide this 6 V minimum at full current, the off-load voltage may be nearer 8–10 V, depending on current. The ripple that is present should also be removed by the stabilizer, and this requires that the negative peak of ripple voltage should not take the

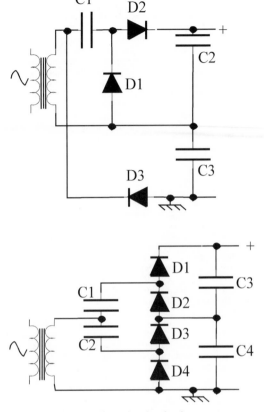

Figure 4.7 Tripler and quadrupler circuits.

output of the rectifier-reservoir system below the minimum voltage for the stabilizer. As ripple is usually quoted in RMS terms, it is not always easy to find out what the minimum transient voltage output from the reservoir capacitor will be.

All of this leads to the use of fairly generous voltage levels at the input to the stabilizer, and this in turn can present dissipation problems. IC stabilizers will have a maximum input voltage stipulated, and this can be surprisingly high, as high as 35 V for a 5 V stabilizer. This voltage difference is across the stabilizer, however, and the stabilizer will have to dissipate power equal to the voltage excess multiplied by current flowing. For example, if a supply into a 5 V stabilizer is 15 V at a current of 1 A, then the excess voltage is 10 V and the power dissipated is 10 W. This is just one of the factors that has led to the development of switch-mode power supplies, see later.

A stabilizer may be connected in series with the current or in parallel with the voltage of a power supply, and in practice the series type is much more common (Figure 4.8). The principle is that the resistance of the non-linear component becomes less as current increases, so compensating for dropping voltage of the reservoir by having a lower drop across this resistance.

All stabilization requires some form of steady standard voltage, and the usual standard is provided by a Zener diode. The Zener diode, named after the discoverer of its principles, Clarence Zener, is not quite what it seems, because two effects are present, one called avalanche breakdown and the other called Zener breakdown. For most purposes, the differences are mainly of interest to researchers, but when we come to look at temperature effects, the differences become more important.

The important feature of the Zener diode is that it is intended to be used with reverse bias, although it possesses a normal forward characteristic like any other silicon diode. The normal reverse characteristic of any silicon diode is to withstand reverse voltages until a threshold value at which reverse current increases rapidly. The Zener type of diode is constructed so that the breakdown of resistance for reverse voltage is much more rapid, leading to a characteristic of the type illustrated in Figure 4.9, with a very steep slope at the breakdown voltage. This implies that over a large range of reverse currents, the reverse voltage will be almost constant, so that this type of diode can be used as a constant-voltage source.

Zener diodes are usually selected so as to offer a set of breakdown voltage levels that follows the normal tolerance sequence, and using values that are written following the convention of BS 1852. This uses the letter V in place of the decimal point so as to avoid any confusion due to misplaced points or to the use of a comma in place of a point, so that you will see Zener diode voltages quoted in terms of numbers such as 4V7, 5V6, 6V8, 7V5 and so on. The effectiveness of a Zener diode for providing a stable output reference voltage is measured by its *slope resistance*, the equivalent of internal resistance. This is not a constant, but something that depends on doping levels, and is not the same for diodes of different power dissipation ratings. For example, its value is smallest for diodes of 0.5 W dissipation whose doping corresponds to a voltage breakdown level of between 6 V and 8 V.

A typical circuit for voltage reference is illustrated in Figure 4.10. The Zener diode is connected in series with a resistor which is used to limit the current. If the supply voltage input varies, but does not fall as low as the Zener voltage, then the current through the diode will

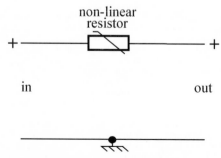

Figure 4.8   The series stabilizer principle.

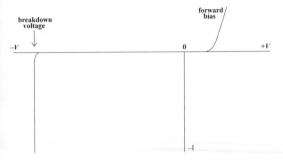

Figure 4.9 Typical Zener diode characteristic.

vary but the voltage across the diode will be almost constant. The current through the resistor is given by $\frac{E - V}{R}$ where E is the supply voltage and V is the Zener breakdown voltage.

If the current level drops to a very low value, usually considerably less than 1 mA, or the voltage level drops to near the Zener level, the breakdown action will cease and the diode becomes non-conducting. This also means that the voltage is no longer held at the reference value, and any stabilization based on this voltage will cease.

The breakdown voltage is temperature dependent, but for one voltage breakdown level of around 4 V the temperature coefficient is almost zero. Below this level, the temperature coefficient is negative, and above this level the temperature coefficient is positive. This effect arises because the observed temperature coefficient is the sum of two separate temperature effects which act in opposite directions and cancel each other out for a breakdown voltage of around 4 V.

The effect of temperature coefficient is to change the voltage breakdown level as the temperature changes. For Zener diodes, the temperature coefficient is usually quoted as % change per °C, so that a typical value for a 7V5 Zener is +0.065%/°C. Using this as an example, if the normal running temperature is 25°C and the Zener is used at 75°C, then the temperature difference is 50°C and this will

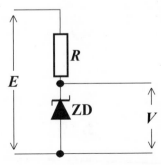

Figure 4.10 A simple Zener diode stabilizer for small currents.

cause a change in voltage of 0.065% × 50 = +3.25%. For a 7V5 Zener, a change of +3.5% is 7.5 × 3.5/100 = +0.2625 V, making the voltage about 7.76. If the temperature coefficient is negative, as it is for values below 4V0, then the change is also negative, meaning that the voltage level will drop as the temperature is raised.

Zener diodes are supplied in a range of breakdown voltages, typically 2V7 to 62V, and in a large variety of power ratings. Small Zeners are rated at around 0.5 to 1.3 W, subject to a derating of typically 9 mW per degree above a temperature of 25°C. Zeners of 20 W rated dissipation are available, but there is usually little point in using large wattage Zener diodes in simple stabilizer circuits, because voltage stabilization can be accomplished more cheaply and efficiently at the same power levels by using a stabilizer IC.

The most basic form of discrete-series stabilizer is illustrated in Figure 4.11. The Zener diode is connected through a resistor to the unstabilized supply voltage, and the diode voltage is applied to the base of the power transistor. The resistor from the output of the stabilized supply at the emitter of the transistor is not essential, but if no other load is connected it helps to establish correct voltage levels quickly when the circuit is switched on. The output of this circuit is not identical to the Zener output, and the variation is rather greater. The Zener diode establishes the base voltage for the transistor, but the voltage at the emitter will be about 0.6 V less, and the difference will become greater as the load current increases. The transistor will usually be mounted on a heatsink, and the current to the Zener diode must be adequate to provide for the base current that the transistor will need when maximum load current is passing between collector and emitter.

The transistor dissipates power only as and when the load dissipates. The load is stabilized against both input voltage variations and load current variations, although the stabilization against load current variation is only as much as can be provided by the use of the transistor in this emitter-follower circuit. The maximum variation of voltage will be equal to the variation of base to emitter voltage for the range of current that is being used. This may amount to no more than about 0.2 V.

For a better standard of stabilization, discrete circuits may be used, but the predominant solution is the IC stabilizer. This does not mean, however, that all voltage stabilization problems can be solved by connecting an IC stabilizer into an existing power supply, and some

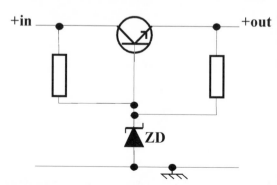

Figure 4.11 A simple discrete stabilizer.

appreciation of the action of the circuitry helps considerably in understanding what is involved in the use of the IC. In particular, since the IC incorporates power transistor, comparator, Zener diode and resistors into one unit, the Zener will be subject to higher temperatures than would usually be encountered in a discrete circuit. The use of heatsinks and calculations of temperature will therefore form an important part of the design of a power supply that uses this type of stabilizer.

IC regulators are available in a bewildering variety of types, but the simplest are the familiar fixed voltage types such as the 7805. The final two digits of this type number indicate the stabilized output, 5 V for this example. The 7805 is a three-pin regulator which requires a minimum voltage input of 7.5 V to sustain stabilization, with an absolute maximum input voltage of 35 V. The maximum load current is 1 A and the regulation against input changes is typically 3–7 mV for a variation of input between 7.1 V and 25 V. The regulation against load changes is of the order of 10mV for a change between 5 mA and 1.5 A load current. The noise voltage in the band from 10 Hz to 100 kHz is 40–50 µV, and the ripple rejection is around 70 dB. Maximum junction temperature is 125°C, and the thermal resistance from junction to case is 5°C/W.

This type of stabilizer, used extensively for power supplies in digital equipment, provides a useful source of examples that illustrate the way that stabilized supplies can be designed around an IC. The recommended circuit is shown in Figure 4.12 with the diode bridge and reservoir capacitor. The capacitors that are shown connected each side of the IC are very important for suppressing oscillations and must not be omitted. In particular, the 330 nF capacitor at the input must be wired across the shortest possible path at the pins of the IC. The maximum allowable dissipation is 20 W, assuming an infinite heatsink, and the actual capabilities are determined by the amount of heatsinking that is used. If no heatsink is used, the thermal resistance of the 7805 is about 50°C/W, and for a maximum junction temperature of 150°C this gives an absolute limit of about 2.5 W, which would allow only 2.5 V across the IC at rated current, an amount only just above the absolute minimum.

The thermal resistance, junction to case, is 4°C/W, and for most purposes, the IC would be mounted on to a 4°C/W heatsink, making the total thermal resistance 8°C/W. This would permit a dissipation of about 15.6 W, which allows up to 15 V or so to be across the IC at the rated 1 A current. This is considerably more useful, since a 5 V supply will generally be provided from a 9 V transformer winding whose peak voltage is 12.6 V, making it impossible to cause over-dissipation at the rated 1 A current since the voltage output from the reservoir capacitor will be well below 12.6 V when 1 A is being drawn.

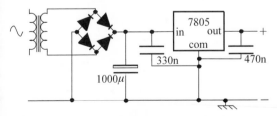

Figure 4.12 The usual IC stabilizer supply circuit.

The 78 series of stabilizers are complemented by the 79 series, which are intended for stabilization of negative voltages. The circuits that can be used are identical apart from the polarity of diodes and electrolytic capacitors, and the range of currents is substantially the same as for the 78 series.

CURRENT RANGES

In general, if higher currents are needed than can be supplied using a stabilizer such as the 7805, then the usual solution is to use a stabilizer IC which is rated for higher currents, or a hybrid stabilizer such as the 78H05 which consists of a power transistor combined with a stabilizer circuit for currents up to 5 A.

The alternative is to use the stabilizer to control a discrete power transistor which is rated to pass the required current. The power transistor will have to be mounted on a large heatsink, preferably not the one that is used for the stabilizer, and since the stabilizer supplies only the base of the transistor it should usually be possible to dispense with a heatsink for the stabilizer itself. Figure 4.13 shows a typical circuit.

PROTECTION CIRCUITRY

In the early days of IC stabilizers, the failure rate of stabilizer chips was very high, but later versions have added protective circuitry which has almost eliminated failures of the type that caused so many problems at one time. The main protective measures are for thermal protection and foldback overload protection. Some knowledge of what goes on inside the chip can be useful, particularly to explain why an unexpected failure happened.

The universal stabilizing circuit makes use of a comparator amplifier controlling a power transistor, and all protection circuits operate on the comparator amplifier circuitry. Thermal protection makes use

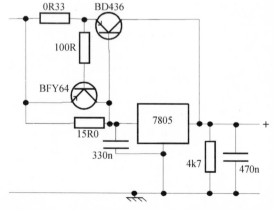

Figure 4.13 A circuit using a power transistor to extend current range. Typical values are shown.

of diodes in the bias path of the amplifier so that the bias is reduced as the temperature increases until the chip passes little or no current. This protects well against long-duration overloads or high ambient temperatures which bring up the temperature of the IC fairly uniformly, but it cannot protect well against transient overloads, and in any circuit which is likely to cause sudden current overloads a reservoir capacitor should be used on the stabilized side in order to supply such transient demands.

Foldback protection is a method of protecting against damage caused by excessive current. Simple circuits which act to protect the stabilizer will only limit the current to its maximum value when the output is short-circuited, (Figure 4.14(a)). The stabilizer will then be passing its full rated current, and the dissipation of the stabilizer and of its protection circuits will be large. The better alternative is the 'foldback' characteristic of Figure 4.14(b), in which the current is reduced as the voltage drops to zero because of a short circuit.

Low-dropout stabilizers

The *dropout* of a stabilizer is the minimum voltage difference which must exist between input and output in order to sustain the action of the stabilizer. The dropout for the 78 series of stabilizer ICs is at least 2 V and for some purposes, particularly for stabilizing supplies that are based on secondary batteries, this differential is too large. Low dropout stabilizers allow much lower levels of input voltage, typically down to 5.79 V for a 5 V output stabilizer.

Low-dropout stabilizers were originally developed for the car industry to provide stabilized outputs for microprocessor circuitry, and they would not be used for general-purpose stabilization for mains-powered supplies. Most low-dropout stabilizers feature additional protection against supply reversal, the effects of using jumper leads between batteries, and large voltage transients. Several types also feature inhibit pins, which allow the stabilizer to be switched back on again after it has been switched off by an overload.

CURRENT REGULATION

The voltage stabilizer ICs can also be used for current stabilization, using the basic circuit of Figure 4.15 which can be used for the recharging of nickel-cadmium batteries. The value of the resistor R determines the current, but to this amount must be added the standing current passed through the common connection pin. For example, the 78 series

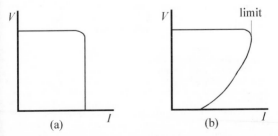

Figure 4.14 Protection methods (a) limiting, (b) foldback.

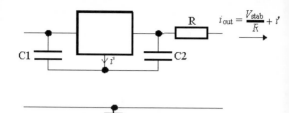

$$i_{out} = \frac{V_{stab}}{R} + i'$$

Figure 4.15 Current stabilization circuit.

of regulators will pass about 4.5 mA through the common connection. The dissipation in the resistor can be considerable, so that this would normally be a vitreous wire-wound type. For a 7812 0.5 A stabilizer, for example, R would be a 24 ohm, 6 W resistor.

SPECIALIZED ICS

There are many more specialized ICs which turn up in servicing, and of these only a few representative types can be mentioned here. The items that are of particular interest are over-voltage protectors and battery backup switching ICs.

Power supplies that are intended to be used with TTL logic circuitry must guard against over-voltage, which can destroy TTL chips very rapidly. The duration of over-voltage that can destroy TTL chips is much too brief to trigger any conventional fuse, so that only other semiconductor circuits can play any useful part in protecting a circuit against the type of failure of a stabilizer that leads to excessive voltage. As it happens, this is the most common type of stabilizer failure, so that the protection is necessary for any TTL circuit of any significance. Many modern digital circuits make extensive use of MOS devices which are less susceptible to damage from over-voltage, but it is unusual to find a large digital circuit which does not contain at least one or more TTL devices

The typical over-voltage protector chip is a sensor which has to be used in conjunction with a thyristor 'crowbar' circuit. A crowbar, as the name suggests, is a short across the output of the power supply which will force fuses to blow and foldback-current circuit in the path of the excess current (if still active) to come into operation. The principle is that the sensor triggers a thyristor which will in the time of a microsecond or so short the power supply output, reducing the voltage and protecting the circuits. Some few milliseconds later, fuses will blow, completing the protective action and ensuring a complete shutdown of the system.

A protective circuit, due to Motorola and using their OVP (over-voltage protection) chip, is illustrated in Figure 4.16. The trip voltage is set by the resistive potential divider, and close-tolerance resistors must be used for R1 and R2, because otherwise the triggering level may be below the normal power supply level, or too high to be useful. The trip voltage level is affected by temperature changes, with a temperature coefficient of 0.08%/°C. The propagation delay of the OVP chip is of the order of 500 ns. Note that the circuitry around the OVP chip should be earthed directly at pins 5 and 7 of the chip rather than to any more remote earth.

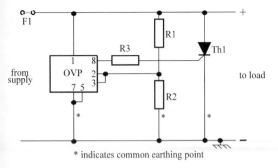

Figure 4.16   Over-voltage trip circuitry.

The resistor R3 controls the current into the gate of the thyristor, and for maximum triggering speed, this resistor should be the minimum possible value that can be used. The value is determined by the amount of current that the OVP chip can supply without damage, so that the value of R3 depends on the voltage level of the normal supply voltage, being about 30R for protecting 20 V. At voltages of less than 10 V the resistor R3 is not needed, and for voltage levels above 35 V, a different circuit is used in any case.

A typical circuit for protecting higher voltage supplies is shown in Figure 4.17. This uses a Zener diode to stabilize the supply to the OVP chip at +10 V, so that no series resistance is needed in the gate connection of the thyristor. The value of R4 for this version of the circuit is chosen so as to pass 25 mA, of which the OVP chip takes a normal steady current of 5 mA. The triggering current of up to 300 mA is provided by the capacitor which must be connected across the Zener diode.

One problem that affects the operation of any OVP circuit is electrical noise. False triggering because of noise transients can be reduced or eliminated by connecting a capacitor from pins 3 and 4 of the OVP to local earth. The value of this capacitor can be in the range 100 pF to 1 μF for delay times in the range 1 μs to 10 ms. A momentary overvoltage will start this capacitor charging and triggering will occur when the voltage reaches the reference level of 2.6 V. If the transient voltage reduces before this level is reached, the capacitor is discharged at about ten times its charging rate so as to reset the delay action for the next transient.

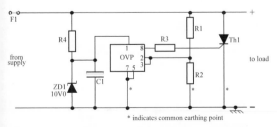

Figure 4.17   Protection circuitry for higher voltage levels.

Two other features of this chip are indicator action and remote triggering. Pin 6 can be connected by way of a load resistor to supply voltage, and the output from this pin can be used to operate a flip-flop (powered from an unstabilized supply) which can be used to shut down the stabilizer. In this way, a stabilizer which uses no foldback limiting, or in which the foldback current is still high, can be shut down so as to reduce the thyristor current, so making it unnecessary to use a heatsink for the thyristor. This also ensures that the fuse will not blow, and since the unstabilized voltage will still be present, this pin can also be used to operate an LED indicator to show that the OVP action has been triggered. The other provision is that a voltage of 2 V or more on pin 5 of the OVP chip will trigger a shutdown whether any over-voltage has been sensed or not. This can be used to ensure a power shutdown when another part of the circuit becomes faulty, so preventing damage that might not otherwise cause over-voltage or over-current conditions.

The comparatively simple type of OVP is likely to be all that is needed for protection of most types of digital circuits, but more elaborate protection chips such as Silicon General's *Power Supply Supervisor* chip are available. This chip will add foldback current limiting if this is not provided for in the stabilizer, along with sensing of over-voltage and under-voltage conditions, indication of type of fault and crowbar protection when used along with a thyristor.

Another type of action is the automatic switching between a mains supply and a battery backup. Simple circuits can make use of a diode in series with the battery supply, biased off when the mains supply is in operation, but a considerably more advanced form of action is obtained when a specialized chip such as that from Harris Semiconductor is used. This will switch completely automatically between mains and battery supplies, connecting the circuit to whichever supply offers the higher voltage level. The input voltage level can range between –0.3 V and +18 V, and the current capability of 38 mA can be extended by using the chip to control power transistors.

Figure 4.18 shows a typical application in which the backup is carried out by a rechargeable battery, trickle-charged by way of a diode and limiting resistor when the mains supply of 5 V is active. On switch-off or failure of the mains supply, the battery supply will be used with the output voltage virtually equal to battery voltage (no 0.6 V diode drop). This lower level can then be used for retaining memory, or for

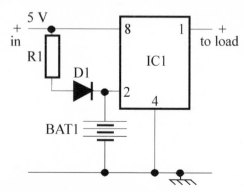

Figure 4.18 Battery backup circuit.

whatever purpose is required. A long-life lithium battery can be used in place of a rechargeable battery, in which case no recharging resistor or diode is needed.

SWITCH-MODE SUPPLIES

The traditional approach to the design and construction of a stabilized supply is in many ways far from ideal. To start with, a large amount of energy is wasted because the unstabilized output level has to be maintained considerably greater than the stabilized level, and the voltage difference will result in considerable dissipation of heat from the stabilizer. At low voltage levels in particular, very large values of capacitance are needed for the reservoir, and this demands electrolytics of 100 000 μF to 500 000 μF, with all of the problems that attach to electrolytics. The low frequency of the ripple from a mains-operated supply is always difficult to remove completely, even using large capacitors, unless the voltage level is high enough to allow inductors with their inevitable series resistance to be used.

The answer to many of the problems of making high-current low-dissipation low-voltage supplies is the use of switch-mode supplies, which exist in several types. The types that are used for applications such as TV receivers use the mains at full voltage to provide power for an oscillator whose output in turn is rectified and stabilized. Another option, used particularly for digital equipment, is to employ a step-down transformer to provide mains isolation, rectifying the output of this transformer to use as the supply to the switch-mode circuits. When this latter approach is used, the circuit can provide for a step down or a step up of the DC voltage applied to the chips, using an inductor when a step up is needed. These circuits also generally make use of higher frequencies for operation.

When full mains voltage is used as the source of the DC for the switch-mode circuit, the isolation of the supply may have to be carefully considered. For TV receiver use it was customary in the past to allow a receiver to have its chassis connected directly to the mains. This eventually fell out of favour because of the need to connect TV through SCART connectors to or from other equipment (such as VCRs or satellite receivers). In addition, most of the voltages that are required for a modern TV receiver are low voltages, and the few supplies that are necessarily of high voltage can be obtained by use of DC–DC conversion techniques. Though most modern designs feature fully isolated power supplies, there are large numbers of old TV receivers still in use which have live chassis working. The form of switch-mode supply for such circuits is the simplest, and an outline of the process is illustrated in Figure 4.19.

In this arrangement the mains voltage is rectified in a bridge circuit, with the negative output of the bridge connected directly to the local chassis earth. This makes it necessary to feed the receiver from an isolating transformer in the course of servicing, and to take great care with the use of instruments which have earthed chassis construction. The positive output from the bridge consists of positive half-cycles at twice mains frequency, and this is used to form the trigger pulses for the switching. Unlike later types of switching circuits, the switching rate is low, so that the need for large smoothing capacitors is not removed, and the main benefit of using the circuit is the elimination of transformers for supplying low voltages. The logic circuitry allows for the regulation of the supply and for foldback of current in the event of a fault condition. The triggering circuits then generate a trigger pulse from the rectified pulses at twice mains frequency and use

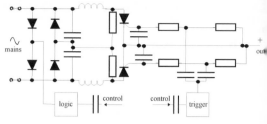

Figure 4.19 Simple form of switch-mode supply.

this to switch the two thyristors. The thyristors operate from the mains input with an output which consists of steep-sided pulses, which can be smoothed and used as a supply at a voltage level of considerably less than the mains supply voltage.

The disadvantage of using low-frequency switching is that smoothing is almost as difficult as it would be for a conventional stabilized supply, but this can be assisted by using *active smoothing*. An active smoothing circuit is similar in outline to that of a series stabilizer, with a reference voltage supplied from a circuit that uses a resistor and capacitor for smoothing. Since this reference voltage needs to supply only a negligible current, the resistor and capacitor can both be of fairly large value, ensuring excellent smoothing of this voltage, and the negative feedback action of the stabilizer circuit ensures that the main supply is also smoothed – you can think of an active smoothing circuit as being a DC amplifier whose input is a perfectly smooth voltage.

The more modern form of switched-mode power supply is illustrated in the block diagram of Figure 4.20. The mains voltage is rectified, using a bridge set of diodes connected directly across the mains, and this output is partially smoothed. The rough DC is used to operate an inverter oscillator whose output is connected to a high-frequency transformer – a typical frequency is 50 kHz so that the transformer can be a comparatively small and light component. All stages up to and including the primary of this transformer are live to mains.

At the secondary of this transformer, low-voltage windings can be used, and these will not be live to mains. The output from a winding can be rectified, using a bridge circuit, and smoothed using comparatively low-value capacitors. For low-current lines, a conventional IC stabilizer circuit can then be used, but more usually a switching circuit will be used, chopping the DC into square waves whose mark-space ratio (ratio of high-voltage time to low-voltage time) can be controlled by a voltage-controlled oscillator. This square waveform is again smoothed into DC, and the level of this DC is used to supply the control voltage for the switching circuit.

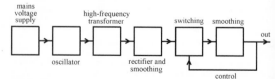

Figure 4.20 Block diagram for modern type of switch-mode supply.

84

In this way, stabilization is carried out with minimal loss of power, because the switching circuits will be either fully conductive or fully cut off, and changes in the stabilization conditions do not cause large changes in dissipation. Heatsink requirements are negligible, and the design of the circuit makes it easy to include cut-off provisions in the event of over-voltage, over-current or overheating. In this form, however, the circuit contains some redundancy in the sense that smoothing is being done twice and the square wave formation is being carried out twice, once in the inverter stage and again in the switching stage. The benefits are precise control of low-voltage high-current supplies, with smaller and lighter components (particularly inductors and capacitors), and with ripple at a frequency which is easily smoothed.

An alternative, still preserving the mains-voltage approach, is to control the inverter part of the circuit which operates at mains voltage. Figure 4.21 shows the block diagram for a circuit which uses an opto-isolator circuit for this purpose. The usual mains-level bridge rectifier and reservoir circuit operates an inverter whose mark-space ratio can be controlled by a DC signal. The output of the inverter is converted to the correct voltage level using a high-frequency transformer as before, and this is rectified and smoothed. The smoothed output is used to control the inverter by way of an opto-isolator acting to supply the DC control voltage from the inverter using as its input the DC output from the stabilized voltage. In this way, the output side of the circuit is not at mains voltage, but can still be used to exert control on the inverter circuit which is at mains voltage.

This approach is comparatively simple in block form, but because of the losses in the opto-isolator it can require rather more circuitry than might appear to be needed. Another problem is that some regulatory bodies concerned with electrical safety do not consider opto-isolators as suitable for total insulation between mains supply and the chassis of electronic equipment. The problem is that most opto-isolators have only a small separation between input and output, less distance than is stipulated to be used between live parts in electrical safety regulations.

The use of pulse-transformer methods is an alternative that can more easily pass electrical safety approval tests, since a transformer can be manufactured to any required standard of isolation. In addition, the use of a pulse-transformer method obviates the need to have an inverter working at mains voltage, and substitutes in its place a pulse amplifier.

An outline block diagram is shown in Figure 4.22. The power supply that is derived from the mains is used as the supply to a pulse amplifier, whose input drive is obtained from the secondary of a pulse

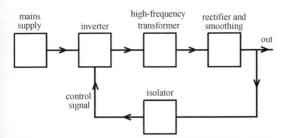

Figure 4.21 A block diagram for a switch-mode circuit using inverter control.

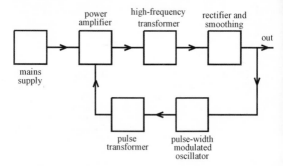

Figure 4.22 An outline block for a pulse-transformer type of switch-mode circuit.

transformer. The output from the pulse amplifier drives the main high-frequency transformer, whose output is at a low voltage and which is rectified and smoothed in the usual way. This voltage is used as the input to a pulse-width modulated oscillator, and the pulses, whose width will be inversely proportional to the output voltage of the supply, are used as driver pulses to the pulse transformer. Although two transformers are used in this circuit, the driver pulse transformer can be a relatively small component, though it must be as stringently isolated between primary and secondary as the main high-frequency transformer. A usual requirement is a long-duration soak test with 3.5 kV or more between windings.

The performance of either type of switch-mode supply can be very impressive. A typical example is the 100 W unit from Weir Electronics (supplied by RS Components) which offers three fully regulated and two semi-regulated outputs. This supply can be obtained in three variants with different output ranges, and the figures quoted are for the HSS 100/2 which supplies +5.1 V at 12 A, +12 V at 5 A and –12 V at 2 A, with subsidiary supplies of –5 V at 1 A and +12 V at 2 A. The +5.1 V 12 A supply is protected against over-voltage, and if the mains supply falls to a level that makes stability impossible to achieve, a power failure signal lamp will illuminate. All outputs are current-limit protected against short circuits, and the level for over-voltage protection is 6.2 V. Line voltage regulation for a 15% line voltage change is 0.1% to 0.5%, depending on the output that is being measured, and the load stabilization for a load change of anything from 20% to 100% is from 2.5% change to 0.25% change, depending on the output measured. At a full 100 W output, the residual mains output ranges from 5 mV to 25 mV, the switching-frequency output from 25 mV to 60 mV, and transient spikes from 50 mV to 120 mV. The unit can be uprated to 150 W if forced-air cooling is available, and there are also other units from the same manufacturer which are rated at 200 W and 300 W respectively with natural cooling.

RF INTERFERENCE

The use of switching-mode supplies and regulators has considerable advantages in weight and size of the main components and in particular the smoothing requirements. One disadvantage, however, is

RF interference. The essence of a switch-mode supply is that large charging and discharging currents are likely to flow, and where these currents flow in stray capacitances and inductances there is likely to be resonance that can result in RF being generated. The problem is tackled firstly in the layout of the circuit by ensuring that earthing, particularly single-point earthing, is correctly carried out, and circuits are encased in metal screens to reduce radiation as far as is practicable, but from then on, filtering is required to reduce the level of conducted RF interference RFI to a minimum.

Filtering is made considerably easier if the circuits use a mains-frequency transformer, since this prevents the spread of RFI through the mains lines. The use of a mains transformer with an earthed screen between primary and secondary, along with a 10 nF capacitor across the output terminals of the transformer, results in a reduction of conducted RFI to well below the limits imposed by regulations of bodies such as the FCC in the USA. Further reductions can be achieved by using an inductive filter on the secondary side of the mains transformer, and by filtering each output of the power supply. The usual range for testing for RFI is 10 kHz to 30 MHz, but good filtering will ensure that low levels of RFI are found for frequencies well above the 30 MHz limit.

SWITCH-MODE SUPPLIES FOR TELEVISION RECEIVERS

The switch-mode power supply systems for TV receivers are constructed in two electrically isolated but coupled sections and complicated by the fact that they are invariably linked to and synchronized by the line timebase circuit. These two sections each have separate earth connections to add further to the problems of faultfinding. This makes it essential that servicing should be carried out using a mains isolating transformer.

Figure 4.23 shows a typical primary mains input section with an earth line, often described as *hot*, that is at half the AC supply line voltage. The circuit is normally fused as shown and uses a full-wave bridge rectifier circuit. The transformer T1 acts as an interference choke to reject mains-borne noise and at the same time prevent the receiver from injecting its own noise back into the mains supply. The unrectified AC is tapped off to provide for the normal degaussing operations at switch-on and the full-wave bridge circuit provides typically 250–300 volts DC output.

This part of the circuit is generally very reliable but the diodes in the bridge can lead to some interesting problems. If one diode develops an open circuit the system reverts to half-wave rectification. This causes the ripple frequency to fall to 50 Hz, the DC output voltage to drop significantly and the ripple amplitude to almost double, resulting in a

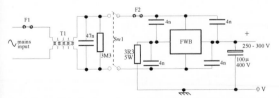

Figure 4.23 A hot-chassis form of switch-mode circuit.

gain reduction and increased hum level that particularly affects the audio and frame amplifier stages. By comparison, if a diode develops a short circuit, the increased current that flows through it will inevitably cause its series companion to fail in a similar way and completely short circuit the bridge network.

The high-value electrolytic capacitor across the DC output can dry out with age causing the DC level to fall, thus affecting any circuit driven from this point.

Until recently, many of the switched-mode power supplies used in TV receivers were in the main constructed from discrete components. These are now progressively being superseded by systems based on integrated control circuits similar to that shown in Figure 4.24. The circuit is powered from a primary supply similar to that shown in Figure 4.23 with which it shares a common earth. Transistor Tr1 acts as the switching or chopper device, whilst T1 performs the function of the chopper transformer and also provides the isolation between this stage and the rest of the receiver circuitry which operates with a separate earth line. These two earth lines should not be bridged under any circumstances. Also when taking voltage measurements, it is important to recognize the appropriate earth line. The voltages of the secondary side of this circuit arrangement are measured with respect to a cold earth. Any signals required in the control of this IC are usually coupled across this barrier, either via opto-couplers or loosely coupled pulse transformers.

The TDA 8380A chip shown can be used in SMPS circuits that operate at any frequency in the range 10 kHz to 100 kHz, using the variable duty cycle concept linked to an RC oscillator whose frequency is controlled by R6 and C4. Connected as shown in Figure 4.24 the circuit operates in the free-running mode. For synchronized use in a TV receiver, pin 11 is connected via an opto-coupler or pulse transformer that is loosely coupled to the line output stage.

The slow or soft start feature operates in two stages; at first pin 5 (V_{cc}) is supplied via the voltage developed across C2 as it charges from a high value resistance R1. Typically the RC values are chosen to provide a delay of about 1 second to protect the chip from switch-on stresses. The second stage of slow start is provided via pin 12 (C3) which provides a controlled increase in the duty cycle. The start of the sawtooth is delayed until this voltage reaches 1.4 V. This slow start rate can also be modified by connecting a resistor in parallel with C3. After the initial stage, V_{cc} is provided via D1 from winding W1 on the chopper transformer. If the supply voltage falls below some threshold minimum defined by a resistor wired between pins 4 and 14 (ground), the circuit switches off and the start-up procedure is repeated. Thus under fault conditions, the circuit may oscillate between these two states.

A band-gap based stabilized reference level of 7 volts is generated in the reference block to power the major sections of the chip. The only parts driven directly from V_{cc} are the initialization stage, the output stages and the series stabilizer transistor. The value of R6 connected to pin 6 determines a reference current which defines a number of parameters. Part of this current is used to charge the sawtooth oscillator capacitor C4 so that the charging time is proportional to the product R6 C4. The maximum duty cycle is set by a resistor connected to pin 12 (R12) and defined by R6/R12. The minimum voltage at pin 4 is proportional to the ratio of R6/R12. The oscillator waveform therefore has a positive going ramp and a negative going flyback.

The lowest of the three inputs to the *controlled slicing level* (CSL), the error amplifier output, the slow start voltage (pin 12) and the *transfer characteristic generator* (TCG) controls the level of slicing. This

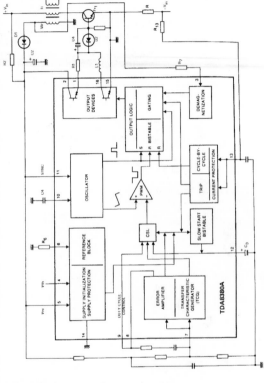

Figure 4.24 Simplified block diagram of SMPS controller chip and circuit A. (Courtesy of Philips semiconductors).

89

stage in conjunction with the *pulse width modulator* (PWM) produces the correct pulse width to control the DC output level. The output pulse thus starts with the beginning of the ramp and ends when its voltage level exceeds that of the slice level. In this way, the variable width output pulses are used via the gating logic to drive the two NPN Darlington switching stages. The bistable circuit is provided to prevent the switching action from being triggered by noise, while the gating logic provides a means of managing the over-voltage and current condition signals.

Under normal operating conditions the voltage supplied to pin 5 (V_{cc}) depends upon the load conditions, if this demands more current, then V_{cc} will tend to fall. A feedback voltage is obtained from V_{cc} via the potential divider network associated with pin 7 and input to the error amplifier. Thus if V_{cc} falls, the corresponding input to CSL will also fall and delay the time before the oscillator sawtooth exceeds the slicing level. This will increase the pulse width so that the switching action creates a rise in the output voltage to meet the demand of the load. The TCG circuit is included to ensure that the error amplifier characteristics are maintained constant over a wide range of input levels.

Because the circuit operates in a non-continuous mode, the chopper transformer may become saturated. This is avoided by coupling W1 via R7 to a demagnetizing circuit when the switch-on action of the chopper transistor is delayed until the transformer current has fallen to zero. R1 and R13 perform part of the function of over-current protection.

The circuit can be synchronized by pulse feedback to pin 11 from the line output stage. The pulse repetition frequency should be somewhat higher than the free-running rate of the sawtooth oscillator which will then be triggered from the negative going edge of the sync pulse.

As shown in Figure 4.24 the circuit is configured to drive a bipolar chopper transistor which is driven between saturation and cut-off. The output devices can source a current of 0.75 A at pin 1 and sink a current of 2.5 A at pin 15. The sourced current which is limited by the value of R1 is used to drive Tr1 hard on for the duration of the major part of the duty cycle. To turn the transistor off, the reverse drive device sinks current through pin 16 to remove the charge on the base of Tr1. L1 is used to limit the rate of change of this value to a safe value.

By making minor changes to the circuit at the output stage, this chip can also be used to drive an MOS power switcher device.

Dummy load testing

Because of the interdependence between the chopper and line output stages it may not be easily apparent whether a fault is caused by one or the other stages. In this case, the following method can often be used with advantage. Disconnect the feed to the line output stage (LOPT) and connect a dummy load in the form of a 60 W lamp across the power supply output. If the supply now produces the correct HT DC level, then the fault must be in the LOPT stage. Even if the fault is found to be in the power section, it is still a good idea to run it into the dummy load until the fault is cleared. Furthermore, the use of a Variac in place of direct connections to the mains will allow the AC input to be increased slowly to avoid sudden overloads that result in fuse destruction.

An incandescent lamp has a positive temperature coefficient, low resistance when cold and high when hot. If a 60 W lamp is connected in series with the AC supply it can prevent the blowing of fuses, diodes and even chopper transistors. If there is a short circuit within the power supply, the lamp will light and then increase its resistance to protect the circuit. As a check on the degaussing circuits, the lamp will light brightly at first and then dim as the positive varistors heat up.

HEATSINKS

Heatsinks are a very important part of power supplies and of power output stages. In general, servicing is concerned with maintenance of existing heatsinks rather than replacement or construction, but some notes on heatsink theory and practice may be useful.

In all heatsink calculations, the thermal resistance is very important, and it is a particularly useful quantity because it so closely resembles electrical resistance, allowing us to imagine all the items in the flow path of heat as resistors in a circuit. The symbol used for thermal resistance is θ (Greek theta), and its definition is given by:

$$\theta = \frac{\text{temperature change in °C}}{\text{power transferred in watts}}$$

This emphasizes that when heat flows, there is always a temperature difference or temperature gradient, and the direction of heat flow is always from high temperature to lower temperature. The transfer of heat from the collector junction of a transistor (or from junctions within an IC) is by conduction, and for conductive heat transfer the thermal resistance of each part of the circuit depends only on the materials and the way that they are joined, not on temperature levels. Convective heat transfer through gases (including air) and liquids is not so simple, nor is radiative heat transfer across a gas or a vacuum.

The path from a collector junction, where electrical power is converted to heat, can be imagined as a set of thermal resistances in series, and, like electrical resistance in series, the quantities add to give a total. For example, if a transistor has a thermal resistance θ_{JC} from junction to case, and there is a thermal resistance of value θ_{CS} from the case to the heatsink, and θ_{SA} from the heatsink to the surrounding air (ambient), then the total thermal resistance in this 'circuit' is $\theta_{JC} + \theta_{CS} + \theta_{SA}$. Multiplying this quantity by the amount of power that is dissipated will give the temperature difference between the ambient air and the semiconductor junction, and adding the ambient temperature (often assumed as 50°C) to this gives the temperature of the semiconductor junction. As an equation, this is:

$$T_j = T_A + P \left(\theta_{JC} + \theta_{CS} + \theta_{SA} \right)$$

The junction temperature calculated from this equation must not exceed the maximum temperature stipulated by the manufacturer.

Very often, you need to find what the maximum value of thermal resistance θ_{SA} can be used, given the assumed ambient temperature and maximum semiconductor junction temperature, with suitable values of the other thermal resistances. In such a case, the equation can be recast as:

$$\theta_{SA} = \frac{T_J - T_A}{P} - \theta_{JC} - \theta_{CS}$$

The θ_{JC} value can be obtained from the specification of the semiconductor, and case to heatsink value, using a mica washer and heatsink grease, is usually of the order of 0.50°C/W.

For example, suppose that the junction temperature is not to exceed

90°C for a power dissipation of 10 W in air at 50°C. The thermal resistance from junction to case is specified by the manufacturer as 1.5°C/W, and the mica washer and grease contributes 0.50°C/W. The maximum possible heatsink thermal resistance is then:

$$\theta_{SA} = \frac{90 - 50}{10} - 1.5 - 0.5 = 2$$

so that the heatsink has a thermal resistance of less than 2°C/W. Note that if you calculate a value that is zero or negative, then no heatsink can possibly be good enough, and the power level must be reduced.

BATTERIES AND THEIR PROBLEMS

Batteries were the original source of DC, and have always been an important form of power supply for electronic equipment. Strictly speaking, a battery is an assembly of single cells, so that the action of a cell is the subject of this section. Any type of cell depends on a chemical action which is usually between a solid (the cathode plate) and a liquid, the electrolyte. The use of liquids makes cells less portable, and the trend for many years has been to jellified liquids, and also materials that are not strong acids or alkalis. The voltage that is obtained from any cell depends on the amount of energy liberated in the chemical reaction, but only a limited number of chemical reactions can be used in this way, and for most of them, the energy that is liberated corresponds to a voltage of between 0.8 and 2.3 V per cell with one notable exception, the lithium cell. This range of voltage represents the range of fundamental chemical actions that cannot be circumvented by refining the mechanical or electrical design of the cell.

The current that can be obtained from a cell is, by contrast, determined by the area of the conducting plates and the resistance of the electrolyte material, so that there is a relationship between physical size and current capability. The limit to this is purely practical, because if the cell is being used for a portable piece of equipment, a very large cell makes the equipment less portable and therefore less useful.

Hundreds of types of cells have been invented and constructed since 1790, and most of them have been forgotten. By the middle of this century, only one type of cell was commonly available, the Leclanché cell (or carbon-zinc cell) which is the familiar type of 'ordinary' torch cell. The introduction of semiconductor electronics, however, has revolutionized the cell and battery industry, and the requirements for specialized cells to use in situations calling for high current, long shelf life or miniature construction have resulted in the development and construction of cells from materials that would have been considered decidedly exotic in the earlier part of the century.

PRIMARY AND SECONDARY CELLS

A primary cell is one in which the chemical reaction is not chemically reversible. Once the cell is exhausted, because the electrolyte has dissolved all of the cathode material or because some other chemical is exhausted, then recharging to the original state of the cell is impossible, though for some types of primary cell, a very limited extension of life can be achieved by careful recharging. In general, simple attempts to recharge a primary cell will usually result in the internal liberation of gases which will eventually burst explosively through the case of the cell, but specialized rechargers that incorporate monitoring of voltage and current can be used to recharge any common type of cell other than the lithium cell.

A secondary cell is one in which the chemical reaction is one that is truly reversible. Without getting into too much detail about what exactly constitutes reversibility, truly reversible chemical reactions are not particularly common, and it is much more rarely that such a reaction can be used to construct a cell, so that there is not the large range of cells of the secondary type such as exists for primary cells. Even the nickel-cadmium secondary cell which is used so extensively nowadays in the form of rechargeable batteries is a development of an old design, the nickel-iron cell due to Edison in the latter years of the nineteenth century. A truly reversible cell needs a recharging voltage that is only slightly above the normal output level.

The important parameters for any type of cell are its open-circuit voltage (the *EMF*), its 'typical' internal resistance value, its shelf-life, active life and energy content. The EMF and internal resistance principles have been mentioned already, and shelf life indicates how long a cell can be stored, usually at a temperature not exceeding 25°C, before the amount of internal chemical action seriously decreases the useful life. The active life is less easy to define, because it depends on the current drain, and it is usual to quote several figures of active life for various average current drain values. The energy content is defined as EMF × current × active life, and will usually be calculated from the most favourable product of current and time. The energy content is more affected by the type of chemical reaction and the weight of the active materials than by details of design.

Cell types

The carbon-zinc dry cell, as the Leclanché type is more often called now, fails either when the zinc casing is perforated or when the chemicals inside are exhausted. One of the weaknesses of the original design is that the zinc forms the casing for the cell, so that when the zinc becomes perforated, the electrolyte can leak out, and countless users of dry cells will have had the experience of opening a torch or a transistor radio battery compartment to find the usual sticky mess left by leaking cells. The term 'dry cell' never seems quite appropriate in these circumstances.

This led to the development of leakproof cells with a steel liner surrounding the zinc. Leakproofing in this way allowed a much thinner zinc shell to be used, so cutting the cost of the cell (though it could be sold at a higher price because of the leakproofing) and allowing the cell to be used until a much greater amount of the zinc had been dissolved. Leakproofing is not foolproof, and even the steel shell can be perforated in the course of time, or the seals can fail and allow electrolyte to spill out. Nevertheless, the use of the steel liner has considerably improved the life of battery-operated equipment.

A different group of cell types makes use of alkaline rather than acid chemicals, so that though the principle of a metal dissolving in a solution and releasing electrons still holds good, the detailed chemistry of the reaction is quite different. The alkaline reactions do not generate gas, and this allows the cells to be much more thoroughly sealed than the zinc-carbon type. Any attempt to recharge these cells with conventional equipment, however, will generate gas and the pressure will build up until the container fractures explosively. Specialized rechargers can, however, be obtained.

The manganese alkaline cell was invented by Sam Ruben in the USA in 1939 and was used experimentally in some war-time equipment, but full-scale production did not start until the 1960s. The EMF of a fresh cell is 1.5 V, and the initial EMF is maintained almost unchanged for practically the whole of the life of the cell. The energy content, weight for weight, is higher than that of the carbon-zinc cell by a factor of 5–10, and the shelf life is very much longer. All of this makes these cells very suitable for electronics use, particularly for equipment that has fairly long inactive periods followed by large current demand. Incidentally, though the cells use alkali rather than acid, potassium hydroxide is a caustic material which will dissolve the skin and is extremely dangerous to the eyes. An alkaline cell must never be opened, nor should any attempt ever be made to recharge it except with suitable equipment.

The miniature cells are the types specified for deaf-aids, calculators, cameras and watches, but they are quite often found in other applications, such as for backup of memory in computing applications and for 'smart-card' units in which a credit card is equipped with a complete microprocessor and memory structure so that it keeps track of transactions. The main miniature cells are silver-oxide and mercury, but the term mercury cell can be misleading, because metallic mercury is not involved.

The mercuric oxide button cell can be classed as an alkaline type. The EMF of such cells is low, 1.2 – 1.3 V, and the energy content is high, with long shelf life due to the absence of local action. The silver-oxide cell is constructed in very much the same way as the mercuric oxide cell, but using silver (I) oxide mixed with graphite as the anode. The EMF is 1.5 V, a value which is maintained at a steady level for most of the long life of the cell. The energy content is high and the shelf life long.

All of these miniature cells are intended for very low-current applications, so that great care should be taken to avoid accidental discharge paths. If the cells are touched by hand, this will leave a film of perspiration which is sufficiently conductive to shorten the life of the cell drastically. When these cells are fitted, they should be moved and fitted with tweezers, preferably plastic tweezers, or with dry rubber gloves if you need to use your hands. These cells should not be recharged, nor disposed of in a fire. The mercury type is particularly hazardous if mercury compounds are released, and they should be returned to the manufacturer for correct disposal if this is possible, or disposed of by a firm that is competent to handle mercury compounds. No attempt should be made to recharge miniature cells, even using a microprocessor-controlled charger.

Lithium cells

Lithium is a metal akin to potassium and sodium which is highly reactive, so much so that it cannot be exposed to air and reacts with explosive violence with water. A lithium cell must never under any circumstance be cut open. The reactive nature of lithium metal means that a water solution cannot be used as the electrolyte, and much research has gone into finding liquids which ionize to some extent but which do not react excessively with lithium. A sulphur-chlorine compound, thionyl chloride, is used, with enough dissolved lithium salts to make the amount of ionization sufficient for the conductivity that is needed. The whole cell is very carefully sealed.

The cell has an exceptionally high EMF of 3.7 V, a very long shelf life of ten years or more, and high energy content. The EMF is almost constant over the life of the cell, and the internal resistance can be low. Lithium cells are expensive, but their unique characteristics have led to them being used in automatic cameras where focusing, film wind, shutter action, exposure and flash are all dependent on one battery, usually a two-cell lithium type.

For electronics applications, lithium cells are used mainly in computers for memory backup, and very often the life of the battery is as great as the expected lifetime of the memory board or motherboard itself. The cells are sealed, but since excessive current drain can cause a build-up of hydrogen gas, a 'safety-valve' is incorporated in the form of a thin section of container wall which will blow out in the event of excess pressure. Since this will allow the atmosphere to reach the lithium, with risk of fire, the cells should be protected from accidental over-current which would cause blow-out.

A typical protection circuit is illustrated in Figure 4.25. This is for use in applications where the lithium cell is used as a backup, so that D1 conducts during normal memory operation and D2 conducts during backup. Short-circuit failure of D2 would cause the lithium cell to be charged by the normal supply, and the resistor R will then limit the current to an amount which the cell manufacturer deems to be safe. If the use of a resistor would cause too great a voltage drop in normal backup use, it could be replaced by a quick-blowing fuse, but this has the disadvantage that it would cause loss of memory when the main supply was switched off.

Lithium cells must *never* be connected in parallel, and even series connection is discouraged and limited to a maximum of two cells. The cells are designed for low-load currents, and Figure 4.26 shows a typical plot of battery voltage, current and life at 20°C. Some varieties of lithium cells exhibit voltage lag, so that the full output voltage is available only after the cell has been on load for a short time – the effect becomes more noticeable as the cell ages. Another oddity is that the capacity of a lithium cell is slightly lower if the cell is *not* mounted with the +ve terminal uppermost.

SECONDARY CELLS

A secondary cell makes use of a reversible chemical process, so that when the cell is discharged, reverse current into the cell will recharge it by restoring the original chemical constitution. Unlike primary cell reactions, reversible reactions of this type (requiring a fairly low supply voltage for recharging) are unusual and only two basic types are known, the lead-acid type and the alkali-metal type, both of which have been used for a considerable time.

The lead-acid cell uses plates made from lead and perforated to allow them to be packed with the active materials. Both plates are immersed

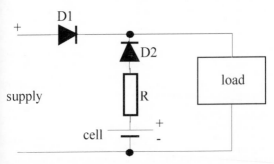

Figure 4.25 Typical lithium cell protection circuit.

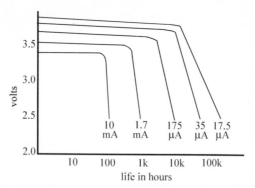

Figure 4.26 Plot of voltage, current and life for a typical lithium cell.

in sulphuric acid solution. The acidity is much greater than that of the electrolytes of any of the acidic dry cells, and very great care must be taken when working with lead-acid cells to avoid any spillage of acid or any charging fault that could cause the acid to boil or to burst out of the casing. In addition, the recharging of a vented lead-acid cell releases hydrogen and oxygen as a highly explosive mixture which will detonate violently if there is any spark nearby. The EMF is 2.2 V (nominally 2.0 V), and the variation in voltage is quite large as the cell discharges.

The older vented type of lead-acid cell is now a rare sight, and modern lead-acid cells are sealed, relying on better control of charging equipment to avoid excessive gas pressure. The dry type of cell uses electrolyte in jelly form so that these cells can be used in any operating position. Cells that use a liquid electrolyte are constructed with porous separator material between the plates so that the electrolyte is absorbed in the separator material, and this allows these cells also to be placed in any operating position. Since gas pressure build-up is still possible if charging circuits fail, cells are equipped with a pressure-operated vent which will reseal when pressure drops again.

Lead-acid cells are used in electronics applications mainly as backup power supplies, as part of uninterruptible power systems, where their large capacities and low internal resistance can be utilized. Capacity is measured in ampere-hours, and sizes of 9 Ah to 110 Ah are commonly used. Care should be taken in selecting suitable types – some types of lead-acid cells will self-discharge considerably faster than others and are better suited to applications where there is a fairly regular charge/discharge cycle than for backup systems in which the battery may be used only on exceptional occasions and charging is also infrequent. Figure 4.27 shows the self-discharge rates of jelly-electrolyte cells at various temperatures, taking the arbitrary figure of 50% capacity as the discharge point.

Lead-acid batteries need to be charged from a constant-voltage source of about 2.3 V per cell at 20°C. Cells can be connected in series for charging provided that all of the cells are of the same type and equally discharged. A suitable multi-cell charger circuit is illustrated in Figure 4.28, courtesy of RS Components. For batteries of more than 24 V (12 cells) the charging should be in 24 V blocks, or a charging system used that will distribute charging so that no single cell is

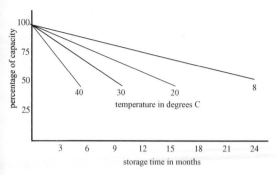

Figure 4.27 Self-discharge rates for lead-acid (jelly type) cells.

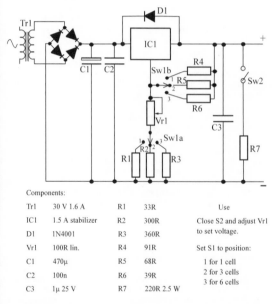

Components:

					Use
Tr1	30 V 1.6 A	R1	33R		
IC1	1.5 A stabilizer	R2	300R		Close S2 and adjust Vr1
D1	1N4001	R3	360R		to set voltage.
Vr1	100R lin.	R4	91R		Set S1 to position:
C1	470μ	R5	68R		1 for 1 cell
C2	100n	R6	39R		2 for 3 cells
C3	1μ 25 V	R7	220R 2.5 W		3 for 6 cells

Figure 4.28 Lead-acid cell charger circuit (courtesy of RS Components).

being over-charged. Parallel charging can be used if the charger can provide enough current.

The operating life of a lead-acid cell is usually measured in terms of the number of charge/discharge cycles, and is greater when the cell is used with fairly high discharge currents – the worst operating conditions are of slow discharge and erratic recharge intervals, the conditions that usually prevail when these cells are used for backup purposes. One condition to avoid is *deep discharge*, when the cell has been left either on load or discharged for a long period. In this state, the terminal voltage falls to 1.6 V or less and the cell is likely to be permanently damaged unless it is immediately recharged at a very low current over a long period. Typical life expectancy for a correctly operated cell is of the order of 750 – 6000 charge/discharge cycles.

Nickel-cadmium cells

The original type of alkaline secondary cell, invented by Edison at the turn of the century, was the nickel-cathode iron-anode type. The EMF is only 1.2 V, but the cell can be left discharged for long periods without harm, and will withstand much heavier charge and discharge cycles than the lead-acid type. Though the nickel-iron secondary cell still exists powering milk floats and fork-lift trucks, it is not used in the smaller sizes because of the superior performance of the nickel-cadmium type of cell which is now the most common type of secondary cell used for cordless appliances and in electronics uses.

Nickel-cadmium cells can be obtained in two main forms, mass plate and sintered plate. The mass-plate type used nickel and cadmium plates made from smooth sheet, the sintered type has plates formed by moulding powdered metal at high temperatures and pressures, making the plates very porous and of much greater surface area. This makes the internal resistance of sintered-plate cells much lower, so that larger discharge currents can be achieved. The mass-plate type, however, has a much lower self-discharge rate and is more suitable for applications in which recharging is not frequent. Typical life expectancy is from 700 to 1000 charge/discharge cycles.

One very considerable advantage of the nickel-cadmium cell is that it can be stored for 5 years or more without deterioration. Though charge will be lost, there is nothing corresponding to the deep discharge state of lead-acid cells which would cause irreversible damage. The only problem that can lead to cell destruction is reverse polarity charging. The cells can be used and charged in any position, and are usually supplied virtually discharged so that they must be fully charged before use. Most nickel-cadmium cell types have a fairly high self-discharge rate, and a cell will on occasions refuse to accept charge until it has been 'reformed' with a brief pulse of high current. Cells are usually sealed but provided with a safety vent in case of incorrect charging.

In use, the nickel-cadmium cell has a maximum EMF of about 1.4 V, 1.2 V nominal, and this EMF of 1.2 V is sustained for most of the discharge time. The time for discharge is usually taken arbitrarily as the time to reach an EMF of 1 V per cell, and Figure 4.29 shows typical voltage-time plots for a variety of discharge rates. These rates are noted in terms of capacity, ranging from one fifth of capacity to five times capacity, when capacity is in ampere-hours and discharge current in amps. For example, if the capacity is 10 Ah, then a C/5 discharge rate means that the discharge current is 2 A.

Charging of nickel-cadmium cells must be done from a constant-*current* source, in contrast to the constant-*voltage* charging of lead-acid types. The normal rate of charge is about one tenth of the Ah rate, so that for a 20 Ah cell, the charge rate would be 2 A. Sintered types can be recharged at faster rates than the mass-plate type, but the mass-plate type can be kept on continuous trickle charge of about 0.01 of capacity (for example, 10 mA for a cell of 1 Ah capacity). At this rate, the cells can be maintained on charge for an extended period after they are fully charged, but this over-charge period is about three times the normal charging time. Equipment such as portable and cordless phones which would otherwise be left on charge over extended intervals

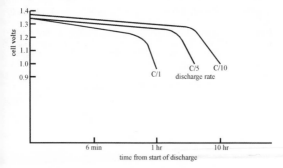

Figure 4.29 Voltage-time plots for nickel-cadmium cells.

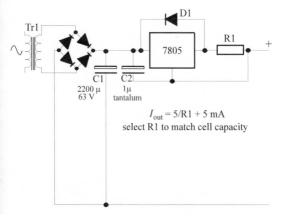

$I_{out} = 5/R1 + 5\ mA$
select R1 to match cell capacity

Figure 4.30 Recharging circuit for NiCd cells, (courtesy of RS Components).

such as Bank Holiday weekends and office holidays should be disconnected from the charger rather than left to trickle charge. This means that a full charge will usually be needed when work resumes, but the life of the cells can be considerably extended if the very long idle periods of charging can be avoided. Another option is to leave the equipment switched on so as to discharge the cells, and fit the mains supply with a timer so that there will periodic recharging.

Figure 4.30 shows a recommended circuit for recharging, courtesy of RS Components. This uses a 7805 regulator to provide a fixed voltage of 5 V across a resistor, so that the value of the current depends on the choice of resistor and not on the voltage of the cell. The value of the resistor has to be chosen to suit the type of cell being recharged; values from 10R to 470R are used depending on the capacity of the cell. Because the regulator system is floating with respect to earth, this can be used for charging single cells or series sets of a few cells. Ready-made chargers are also available which will take various cells singly or in combination, with the correct current regulation for each type of cell. A combined discharger/recharger, as used for camcorder batteries, is the preferred method, ensuring that each cell is discharged before it is charged again. This avoids the 'memory' effect that plagues these cells, causing them to lose capacity if they are frequently charged before being completely discharged. Other forms of nickel cell have been devised, notably the nickel hydride (NiH) cells. These are claimed to eliminate the 'memory' effect, so that a discharger is not needed.

CHAPTER 5

AERIALS, FEEDERS AND PROPAGATION

AERIALS AND SPECIFIC PROPERTIES

The aerial of any communications system is a complementary device in that it is equally efficient in either the transmitting or receiving modes. This is advantageous because it allows the various properties or parameters to be evaluated under either condition as most convenient. For example, the setting up and tuning of an aerial system for transmission can nearly always be achieved most easily in the receive mode, provided of course that the operating frequencies are the same.

The main feature of any aerial in either mode is to match the impedance of free space to that of the feeder cable. Because of this and the fact that radio communications are carried over frequencies ranging from less than 50 kHz to more than 100 GHz, there are many aerial configurations that can be used for any particular frequency range.

All electromagnetic waves propagate through space with two components at right angles to each other (orthogonal). These are the electric and magnetic field vectors with strengths of E volts per metre (V/m) and H amperes per metre (A/m) respectively. The ratio E/H is referred to as the characteristic impedance of free space and is equal to 120π ohms or about 377 Ω. If a power of P watts is applied to an aerial that radiates equally well in all directions (an isotropic radiator), then the electric field strength at a distance d metres is given by:

$$E = \sqrt{(30P)}/d = (5.48\sqrt{P})/d \text{ V/m}$$

However, if the isotropic radiator is replaced by a half-wave ($\lambda/2$) dipole that has a preferred direction of radiation, then the electric field strength rises to $E = (7.014\sqrt{P})/d$ V/m.

Thus the $\lambda/2$ dipole has a gain relative to that of the isotropic radiator in the preferred direction of $7.014/5.48 = 1.279$ or 2.14 dB.

Figure 5.1 shows the voltage and current distribution of standing waves along a $\lambda/2$ dipole and the radiation/reception pattern of radio energy. Figure 5.1(d) also shows one way of matching the feeder impedance to that of the dipole aerial.

The gain G of an aerial can be expressed in one of two ways: either

$$G = \frac{\text{Power supplied to a reference aerial}}{\text{Power supplied to aerial under test}}$$

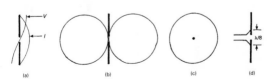

Figure 5.1 Half-wave dipoles: (a) voltage/current distribution; (b) vertical polar diagram; (c) horizontal polar diagram; (d) delta-match.

where the powers in each case produce the same field strength at a given point. Or alternatively as

$$G = \frac{\text{Voltage received by aerial under test}}{\text{Voltage received by reference aerial}}$$

from the same source of electromagnetic energy. This demonstrates just one feature of the complementary nature of these devices. If a λ/2 dipole is exposed to an electromagnetic radio wave with its length parallel to the electric field (described as the plane of polarization of the wave) and at right angles to the magnetic component, it will act as a voltage generator with an impedance of about 75 ohms generating $\lambda E/\pi$ volts.

Now assuming that this aerial is correctly matched to a load, the terminating voltage will be $\lambda E/2\pi$ volts, or −6 dB relative to the electric field strength component (note the λ/π relationship).

The Yagi array

Each aerial has a characteristic feed point or radiation resistance R that depends chiefly upon its dimensions and operating frequency. R is the ratio of the voltage to current on the aerial and is *not* a physical resistance, but more an equivalent one. If the aerial current is I amps, then the radiated power will be I^2R watts. If the aerial is replaced by a real resistor of the same value, then this will radiate an equal amount of power but now in the form of heat. If parasitic elements longer and/or shorter than λ/2 are added in front and behind the dipole as shown in Figure 5.2(a), the gain of the aerial will change as indicated and the radiation pattern will be modified so that its response now has a preferred direction. Although this lowers the dipole impedance, it can be compensated for by making the dimensional changes shown in Figure 5.2(b). This compound aerial is now referred to as a Yagi array. Just like an amplifier, an aerial has a gain–bandwidth product

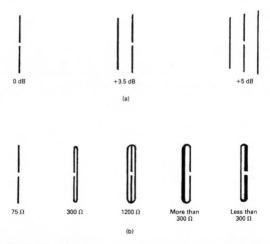

Figure 5.2 Half-wave dipoles: (a) gain relative to dipole; (b) dipole impedance.

that is effectively a constant. Increasing the gain reduces the bandwidth by the same proportions. At the same time, the addition of the shorter director and longer reflector elements reduces the aerial beamwidth as shown in Figure 5.3. The beamwidth being defined as the angle α subtended between the two –3 dB points on the gain-directivity contour.

Figure 5.3 also shows how the addition of the parasitic elements produce minor responses at angles other than the preferred maximum. This leads to a further parameter described as the *front to back* ratio. This is simply the ratio of the gains of a directional aerial at the maximum response from the main lobe to that obtained when the aerial is rotated through 180°.

In practice, the dipole is cut about 8% shorter than the physical $\lambda/2$ length to compensate for the fact that the wave travels slower along the dipole than it does through free space.

SPECIFIC AERIALS

Virtually any piece of metallic conductor that can be isolated from earth so that it responds to either the electric or magnetic field vector can be used as an aerial. Because of this, there are many different types of aerial in use, often designed to suit a particular range of frequencies. These range from ferrite rod aerials used for long and medium waveband receivers; rod, whip or monopole devices employed with car radio receivers; monopole devices that are loaded with ferrite or an inductor to shorten their effective length; dielectric aerials used for portable cellular telephones; Yagi arrays used for many VHF and UHF operations; metal plate aerials used for low to high microwave frequencies and parabolic reflectors.

Ferrite rod aerials

These devices consist of a slab of ferrite material upon which is mounted the tuning and coupling coils. Ferrite material has the effect of concentrating any magnetic field into itself, thus greatly increasing the magnetic coupling between the electromagnetic wave energy and the receiver to which it is connected. To achieve this effect the rod needs to be positioned parallel to the magnetic field vector, so these devices have a figure-of-eight polar response that provides a useful degree of directivity to reject unwanted interference. Ferrite also appears in the aerials of some VHF portable transmitter/receivers. In these cases, the whip aerial is shortened by forming part of it into a coil and then loading the inductive component with a slug of ferrite.

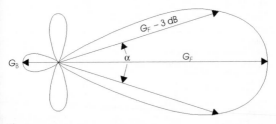

Figure 5.3

The weakness of the ferrite rod lies in its brittleness. If it is dropped it will break and cannot be usefully rejoined with glue because this will effectively isolate the rod into two separate parts.

Rod or monopole aerials

These almost simulate a $\lambda/2$ dipole aerial because the monopole is accompanied by a phantom $\lambda/4$ monopole reflected into the earth plane. However, unlike the dipole, the directivity is such that the maximum response occurs at an angle between the rod and the ground plane.

Dielectric aerials

These consist of small metallic plates supported on a printed circuit type board that functions as the dielectric in a capacitor. The electric field vector then induces a signal via the charging and discharging of the capacitor to power the receiver. In the reverse sense, the alternating current from the transmitter drive creates the radiation via the charging and discharging effect. In practice, these are often used with hand-held mobile phones where their gain and directivity is just as effective as that of the loaded whip version.

Metal plate aerials

These microwave aerials are often referred to as planar arrays because of the way in which they are constructed. Various shaped metal plates are formed in printed circuit manner on a flat substrate, the shapes ranging from rectangular to circular. These are coupled together in a matrix fashion using sections of microwave transmission lines so that the radiation off the patches is in phase in some particular direction. Often the direction of radiation can be steered by electronically varying the feed point position within the matrix. These devices are popular with mobile systems such as those used in aircraft. The panels can be shaped to conform to the fuselage or mounted in aerofoil shaped *shark fins*.

Yagi arrays

These arrays are constructed with typically a maximum of 18 directors, one folded dipole and a reflector structure to provide a gain of up to 20 dBi. The Yagi works well for frequencies from around 10 MHz up to about 1.5 GHz and is commonly used for VHF/FM radio reception, plus VHF and UHF television. For the latter applications, the aerials for either vertical or horizontal polarizations are constructed to operate over a band of channels as indicated in Table 5.1. The Channel 5 TV service is allocated to channels 35 or 37 and this range will be covered by modified Group A or B aerials.

Table 5.1

Channels	Group or band	Colour code
21–34	A	Red
39–53	B	Yellow
48–68	C/D	Green
39–68	E	Brown
21–68	W	Black

Yagi arrays beyond about 20 elements become unwieldy particularly in high winds. In this case, two smaller structures can be coupled in parallel via phasing lines to achieve the same degree of gain and directivity. At the higher UHF frequencies the separate channels, although all emanating from the same transmitter mast, can arrive at a receiving aerial from slightly different directions. The finally selected beam heading then has to be a compromise. The reflected signals arriving over various paths can give rise to a strong and misleading meter reading, only to find that the displayed image contains many ghosts. In such situations, the best type of signal strength instrument is one that provides a monochrome display of the received signal so that the beam heading can be fine tuned to minimize this effect. In an emergency a small portable TV receiver can be pressed into service to act as a very useful signal strength indicator.

THE SERVICE AREA

This is defined as the region around a transmitting aerial where the received field strength is such as to provide a signal quality better than the subjective grades 3 to 4 on the ITU-R (CCIR) 5 point scale as shown in Table 5.2.

The principal impairments that limit the service area are the signal-to-noise (S/N) ratio, multipath interference (TV ghosting), and co-channel interference. For radio the service area is often more accurately defined by the level of man-made noise such as inter-ference from electrical machines and other radio transmissions. For MF (medium frequency) and VHF radio, field strength contours of 3 mV/m and 1 mV/m respectively usually prove satisfactory. Because many listeners employ in-car radio systems, many of the VHF transmissions use slant or circular polarization to minimize the problems for the travelling receiver.

The service area for UHF television requires a field strength in the order of 58 dBµV/m for Band IV (Ch 21–34) and 64 dBµV/m for Band V (Ch 39–68).

The terminating signal voltage at the receiver input depends upon the aerial gain, the feeder loss which is frequency sensitive, the system impedance matching and the aerial pointing accuracy. Typically the TV receiver will provide good quality images when the input terminating signal level lies between 1 and 5 mV. If this level falls to around 200 µV the picture will be noisy and unviewable. Signals above the 5 mV level are likely to overload the receiver AGC (automatic gain control) system to introduce patterning, line tearing or superimposed images. The two main elements of picture quality are thus the electric field strength and the signal input to the RF socket of the receiver which are related as follows.

Table 5.2

Quality	Grade	Impairment
Excellent	5	Imperceptible
Good	4	Perceptible but not annoying
Fair	3	Slightly annoying
Poor	2	Annoying
Bad	1	Very annoying

The field strength is dependent on the transmitter power distribution and the terrain across which the signal propagates. In the case of satellite links, the characteristics of free space are accurately predictable. However, for terrestrial transmissions there is a need to consider such features as the height of the aerial, the location of trees and vegetation that can create anomalies that are seasonal, and multipath effects. It has been found that even a reduction of aerial height can improve the image quality under multipath propagation conditions. It was pointed out earlier that the receiver terminating input voltage falls with a rising frequency. A correction factor in the form of a λ/π ratio has been developed to express this relationship as shown in Table 5.3.

The relationship shown in Table 5.3 can be demonstrated as follows. Assume a Group B aerial with a nominal gain of 11 dB, a feeder loss of 3 dB, a terminating attenuation of 6 dB, and a λ/π correction factor of –16.5 dB. To provide an input of 1 mV or 60 dBμV requires an approximate field strength of 74.5 dBμV/m (5.3 mV/m).

$$11 \text{ dB} - 3 \text{ dB} - 6 \text{ dB} - 16.5 \text{ dB} = -14.5 \text{ dB},$$
$$60 \text{ dB}\mu\text{V} + 14.5 \text{ dB} = 74.5 \text{ dB}$$

For Band II VHF frequencies the λ/π correction factor is approximately equal to unity or 0 dB, so that this factor can be ignored. For Band II the cable loss is lower at typically 1.5 dB in 15 metres. In practice, a high-grade receiver will provide grade 4 or better quality with an input as low as 200 μV. By comparison, a low grade receiver will need an input of around 1 mV to achieve grade 3 quality.

CHANNEL ALLOCATION AND FREQUENCY REUSE

In order to maximize the use of the available and limited frequency spectrum, the UHF television channels are organized in such a way that any given carrier frequency may be reused without creating mutual adjacent or co-channel interference. For example, in the UK the UHF television system with its 44 channels and four channels per broadcast site, means that there can be 11 standard channel groupings. By making careful choices, bearing in mind the propagation conditions, physical separation, transmitter output powers, aerial polar diagram and signal polarization, such interference can be avoided under normal conditions. The channels are therefore generally allocated on the basis of:

$$n, n + 3, n + 6, n + 10,$$
$$n, n + 3, n + 7, n + 10,$$
$$n, n + 4, n + 7, n + 10$$

However, to meet the problems in specific areas the following are also in use:

$$n, n + 3, n + 6, n + 16, \text{ or}$$
$$n, n + 3, n + 15, n + 18$$

Table 5.3

Aerial group	Mid-band frequency	λ/π	dB
A	520 MHz	0.58	–14.7
B	635 MHz	0.48	–16.5
C	770 MHz	0.40	–18

Table 5.4 European broadcasting frequencies

Frequency range	Application
160–225 kHz	Long wave (LW or LF)
525–1605 kHz	Medium wave (MW or MF)
41–68 MHz	Band I VHF (Channel 1–5)
174–216 MHz	Band III VHF (Channel 6–13)
470–582 MHz	Band IV UHF (Channel 21–34)
614–854 MHz	Band V UHF (Channel 39–68)
583–599 MHz	Channel 5 broadcasting (UK only)

In a general sense, channels n ± 1 and n ± 9 may be responsible for adjacent and image channel interference respectively. To further reduce the risks of patterning, the frequencies of the carriers for neighbouring transmitter sites may be offset by ±5/3 or ±10/3 times the line frequency.

It will be noted that Table 5.5 shows a number of conflictions between the US and European designations.

REFLECTOR ANTENNA SYSTEMS

At frequencies above about 1.2 GHz a Yagi array of sufficiently high gain becomes physically unstable and the increasing number of unwanted side lobes in its polar response contours causes it to be less effective against interference signals. At this point, the Yagi gives way to aerials based on the parabolic reflector design, or for specialized applications, the planar patch system.

The mathematical equation for a parabola is $y^2 = 4ax$, as described in Figure 5.4. For such a shape, any ray or wave emanating from point 'a' will reflect off the curve parallel to the x axis. If such a curve is rotated around the x axis, the resulting volume forms a parabolic reflector dish. If this surface is used as a transmitting device, any energy emanating from 'a' will be reflected to form a parallel beam. Conversely, any energy received along a complementary path will be concentrated on to the focal point at 'a'. Furthermore, the total path length from 'a' to the aperture plane via reflection is a constant. Thus when used as a transmitting device, all the energy emanating from 'a' will be reflected and pass through the aperture plane totally in phase. It is this feature that is responsible for the very high forward gain of the dish. The parabolic reflector antenna shape is the most efficient in

Table 5.5 Microwave frequencies (GHz) in common use

Band	Europe	USA
L	0.375–1.5	1.0–2.0
S	1.5–3.75	2.0–4.0
C	3.75–6.0	4.0–8.0
J	11.5–18.0	8.0–12.5
Ku	–	12.5–18.0
K	18.0–30.0	18.0–26.5
Ka	–	26.5–40.0

terms of gain, narrow beamwidth and smallest area. Such a dish can be perforated to reduce windage without losing efficiency, provided that the holes are less than about $\lambda/8$ in diameter.

The mathematical equation for the dish shown in Figure 5.4 can be restated in terms of its diameter D, depth C and focal distance F as follows: $(D/2)^2 = 4FC$ or $C = D/16(F/D)$. This shows that depth depends on the F/D ratio.

If $F/D = 0.25$ the focal point lies on the aperture plane and it becomes difficult to illuminate the whole dish surface efficiently. A high F/D ratio produces a flatter dish which optimizes the forward gain, while a lower ratio gives better side lobe rejection. Typical compromise values thus lie in the range of 0.35–0.45.

The gain of the reflector antenna is given by $G = 4\pi A_e / \lambda^2$, where A_e is the effective area of the aperture plane and typically less than the physical area due to the blocking effect provided by the necessary mountings that support the electronic equipment placed at the focal point. Because A_e is usually large compared with the wavelength of the working frequency, the gain can be very high, typically 35 to 50 dB.

Figure 5.5 shows various methods that can be adopted to minimize the blocking effect. By placing a small sub-reflector at the focal point, the obstruction is smaller and the electronics unit can then be placed in a more convenient position at the centre of the dish.

If a circular section is cut from the outer edge of a larger parabolic reflector, the focal point is no longer in front of the dish as shown by Figure 5.6. The blocking effect has now been removed and consequently the gain has been increased. The offset angle indicated in Figure 5.6 is typically around 28°.

A further variation utilizes an elliptical-shaped reflector with the

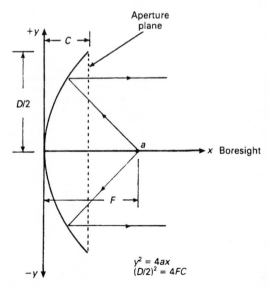

Figure 5.4 The parabolic reflector.

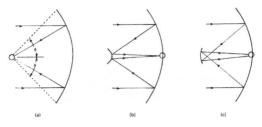

Figure 5.5 Feeding the parabolic reflector antenna: (a) prime focus; (b) Cassegrain sub-reflector; (c) Gregorian sub-reflector.

longer major axis aligned horizontally to increase the gain and reduce beamwidth in this plane. The antenna then becomes more able to discriminate between closely located satellites. The lower gain and wider beamwidth in the vertical direction is not needed to reject signals from interfering satellites, but the side lobes can increase the background noise level. This example of compromise gives portability with reduced size but at the expense of S/N ratio.

FEEDER CABLES

Figure 5.7(a and b) show the general construction of the two most popular cables that are used to transfer signal energy between the aerial

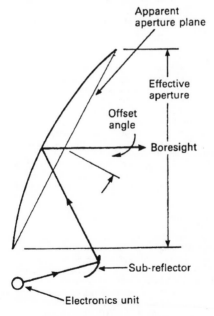

Figure 5.6 Offset feed antenna.

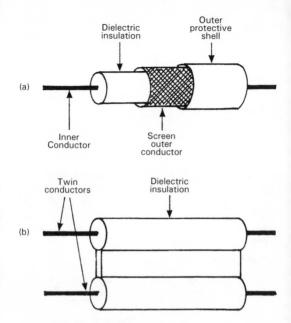

Figure 5.7 Typical transmission line construction: (a) Coaxial cable; (b) Parallel twin feeder cable.

and transmitter or receiver. The important properties should be low loss to the signal, and well-defined and stable characteristic impedance to maintain matching between the aerial and load. The characteristic impedance Z_o is defined as the input impedance of an infinitely long length of cable or the input impedance of a short length when terminated in that value of impedance. In practice, Z_o depends on the dimensions of the conductors, their spacing and the quality of the dielectric material that separates them (the dielectric constant).

In general, coaxial cables (coax) have a $Z_o = 50\,\Omega$ or $75\,\Omega$, while for ribbon or twin feeders, $Z_o = 300\,\Omega$. Both feeders are capable of rejecting interfering signals in the following manner. The outer screen or braid of coax is earthed and thus acts as a shield to the inner signal-carrying conductor. Because only one conductor is directly earthed, this system is described as an unbalanced feeder. The twin feeder is connected to the receiver or transmitter via a centre-tapped transformer winding with the tap point earthed. Signal currents then flow in opposite wires simultaneously in opposite directions. Interfering signals generate in-phase components that flow to earth through the centre tap to produce a self-cancelling effect. Because each wire is balanced to earth, this type of feeder is described as a balanced system. Balanced and unbalanced feeder systems can only be connected together via a balanced to unbalanced (balun) transformer. The coaxial arrangement is commonly used for television systems, while the balanced feeder is often used on the European mainland for VHF television.

Mismatching between the antenna, feeder and terminating device gives rise to losses via reflections. Any signal energy sent along a line to a load will only be totally absorbed if the system is correctly matched. Under mismatch conditions, the load can only absorb an amount of energy dictated by Ohm's law. Any surplus energy is then reflected back along the line so that the forward and reflected waves add and subtract all along the line to produce standing waves, the magnitude of which are indicative of the degree of mismatch. These losses can be quantified by the following parameters and formulae.

The reflection coefficient $(r) = (Z - Z_o)/(Z + Z_o)$, where r is the mismatch impedance.
The Voltage standing wave ratio (VSWR) $= (1 + r)/(1 - r)$.
Also VSWR $= Z/Z_o$, where Z is again the mismatch impedance.
Also $r = (\text{VSWR} - 1)/(\text{VSWR} + 1) \times 100\%$.

Ideally r should be equal to zero so that the VSWR = 1:1. In practice the lowest achievable value is in the order of 1.2:1.

The return loss which represents the ratio of the reflected to incident power at the load may also be quoted as $-20 \log r$ dB.

One unexpected source of mismatch often occurs when a 75 Ω system is inadvertently coupled to a 50 Ω instrument to produce a VSWR value of 1.5:1.

IMPROVING THE SIGNAL STRENGTH LEVEL

It is important to ensure that the aerial, feeder and receiver input impedance matching is maintained throughout the system lifetime. Any metallic components exposed to the weather are likely to corrode and increase the connection losses. It is important to avoid the ingress of water into the feeder system and to this end the cable joints should be securely sealed.

MAST HEAD AMPLIFIERS

Ideally the most effective installation will involve the use of an aerial with suitably high gain coupled to the receiver via low-loss cable. However, in poor signal strength areas there may be a temptation to install an aerial preamplifier but this should only be employed after considering the following points. It must be remembered that the signal received by the aerial already carries a noise component and that the feeder cable will not only create attenuation but it will also add its own noise to degrade the S/N ratio even further. Again, an amplifier also adds its own noise component to the already noisy signal. Thus the amplifier must be fitted close to the aerial and coupled to the receiver via a very high-grade feeder cable that can carry the DC feed from the receiver supply point.

In certain cases of stubborn interference, it might be an advantage to employ two lower gain aerials, stacked in parallel and coupled via suitable phasing lines.

PROPAGATION EFFECTS

Radio horizon

Electromagnetic waves with frequencies above about 30 MHz suffer from some refraction in the atmosphere and hence follow a curved path. Under some circumstances this can extend the radio signal range by about 33% relative to the visual horizon. However, reflections from the earth's surface can create problems due to the energy loss from the wave through scattering and multipath reception. Interference arises when the signal energy arrives at the receiver over several different path lengths. These signal components can either aid or oppose each other and the exact effect depends on the path length differences which in turn create phase differences. At the point of a reflection, the wave undergoes 180° of phase change so that if the path lengths differ by an exact number of wavelengths the signals arrive antiphase and are subtractive. When the path lengths differ by an odd number of half wavelengths the waves arrive in phase and are additive to increase the received signal level. If the point of reflection is moving as in the case of an aircraft, this results in a rapid pulsing/fading effect that is referred to as aircraft flutter. In the case of television, multipath effects give rise to second or subsequent images displaced to the right of the true picture. This so-called ghosting effect may be positive or negative depending on the phase relationship of the signal components.

Unwanted ground reflections can also create multipath effects. The cure in this respect usually involves raising or lowering the aerial height, or moving it to a less affected site.

Knife edge refraction or mountain edge scattering

When a parallel beam of wave energy passes over a knife edge, the beam scatters causing the energy to spread. In a similar way, when a radio wave passes over the top of a mountain edge, some of the wave energy scatters behind the mountain. Thus a distant aerial in the shadow of the mountain can receive an unexpected signal. The phenomenon is related to the Fresnel zone effect and particularly affects signals in the frequency range of 100 MHz to 300 MHz (VHF/UHF). The sharper the mountain edge, then the greater is the degree of refracted energy. The effect is reduced if the mountain top is covered in snow or vegetation. The received signals tend to be restricted to a window area, which is about the same distance from the mountain ridge as the transmitter. There are many recorded instances where signals have been received more than 300 km beyond the mountain.

IONOSPHERIC EFFECTS

While the ionosphere is largely the reason that radio communications are possible at all, this medium has some strange anomalies that create interfering effects that virtually cover the whole of the frequency range involved.

Ionospheric cross modulation
Non-linearities within the ionosphere can cause the modulation of a strong signal to be transferred to a weaker carrier.

Auroral effects
Sporadic intense solar radiation can interact with the earth's magnetic field particularly close to the polar regions to give rise to the well-known visible phenomenon. During such a period, there is considerable interference to radio signals well into the UHF range. If signals are beamed into this region, they tend to be reflected in a north–south direction to cover a much greater range than usual. However, the distortion produced only allows for narrow band signals to be transmitted with any useful reliability.

Inversions

Normally within the atmosphere, the temperature falls with an increase in altitude, but when the inverse occurs it affects the distribution of water vapour. The temperature/humidity feature then becomes horizontally layered to produce sudden changes of refractive index. This causes signals typically in the VHF/UHF range to be carried well beyond the normal horizon, giving rise to co-channel interference. It is possible for several layers to form at different altitudes simultaneously.

Advection inversions

These mostly occur over the sea or low-lying coastal regions and are created by the circulation of a warm dry air mass over the cooler and moist air close to sea surface. With the absence of wind, this condition can last for several days and can cause severe co-channel interference right up into the SHF band. The effect, which can often be predicted a few hours in advance from meteorological data, can also enhance the narrow band communications activity.

Radiation cooling

When a calm cloudless night follows a warm day, the land mass cools fairly quickly. This forms a duct close to the earth's surface that causes enhanced propagation and hence co-channel interference to some services. The problem is normally short lived and is less troublesome when an undulating earth surface exists between transmitter and receiver.

Weather fronts

Often these contain regions of super-refractivity that produce enhanced propagation for VHF/UHF signals which can produce co-channel interference. This is usually associated with a weakening anti-cyclonic system and the interference is often of a short duration.

INTERFERENCE REJECTION

Quarterwave stub

The standing waves produced on a quarter wavelength of transmission line can form the basis of a very effective filter to remove specific forms of interference. At the appropriate lengths, the feeder acts rather like a resonant circuit. An open circuit length of line has a very low impedance just $\lambda/4$ from this end and a very high impedance at $\lambda/2$ distance. By comparison, a short-circuit line behaves in exactly the opposite way. By connecting a short-circuited $\lambda/4$ length of feeder across the aerial input socket any interfering signal at the frequency that makes it such a length will be effectively trapped. The length of the stub is easily calculated, typically about 66% of the free space $\lambda/4$ value for solid coax (85% for air spaced) and then finally tuned by cutting back until maximum attenuation is achieved. Z-or S-shaped patterning on a television screen can arise from faulty receiver RF stages, aerial preamplifiers or even illegal video re-broadcast systems. The radiation from a receiver local oscillator tuned to channel n + 5 can also affect a nearby receiver. This type of problem can also occur with a VCR/satellite system due to the use of remodulation. The usual cure in this case, which may only become obvious during VCR replay, involves retuning the offending modulator to a non-interfering channel.

Toroidal chokes

Any lead to a receiver that can act as a long wire aerial is capable of introducing interference, even the braiding of coax cable and the mains lead can introduce problems. The usual remedy in these cases is to wind a section of the cable in the form of about six turns, through a toroidal ferrite ring. This device is often referred to as a *braid-breaker* and acts as a choke to prevent the interfering signals from reaching the receiver. In stubborn cases, it might be necessary to fit ferrite beads to the aerial input socket or the base lead of the input RF transistor. Again, these devices act like chokes to the interference. Similar problems can arise with Hi-Fi systems that are interconnected with long leads. In this case the cure is simply to use shorter, more direct connecting leads.

CONFEDERATION OF AERIAL INDUSTRIES

This body was set up to help maintain the high standards necessary for all who work in the industry, from the manufacturers to the installers. CAI achieves these ends by representing its members on government, local authority and other national bodies. It also provides information services, training courses and technical advice for its members, thus ensuring that they can provide the highest standard of service to the public. Code of practice booklets which offer sound advice for safe and reliable operations are available from:

Confederation of Aerial Industries
Fulton House Business Centre
Fulton Road
Wembley Park
Middlesex HA9 0TF
Tel. 0181 902 8998, Fax 0181 903 8719

CHAPTER 6

AUDIO SYSTEMS

INTERCONNECTIONS AND SIGNAL LEVELS

Audio connectors originated with the remarkably long-lived jacks which were originally inherited (in the old ¼ inch size) from telephone equipment. Jacks of this size are still manufactured, both in mono (two-pole) and stereo (three-pole) forms, and either chassis mounted or with line sockets. Their use is now confined to professional audio equipment, mainly in the older range, because there are more modern forms of connectors available which have a larger contact area in comparison with their overall size. Smaller versions of the jack connector are still used to a considerable extent, however, particularly in the stereo form. The 3.5 mm size was the original miniature jack, and is still used on some domestic equipment, but the 2.5 mm size has become more common for mono use in particular. The smaller sizes, however, can be a source of trouble when strands of cable break free and cause short circuits.

One of the most common forms of connector for domestic audio is still the phono connector whose name indicates its US origins. Phono connectors are single channel only, but are well screened and offer low-resistance connections, easy soldering, and sturdy construction. The drawback is the number of fittings needed for a two-way stereo connection such as would be used on a stereo recorder. Many users, however, prefer the phono type of plug on the grounds of lower contact resistance and more secure connections.

The European DIN (Deutsches Industrie Normallschaft, the German standardizing body) connectors use a common shell size for a large range of connections from the loudspeaker two-pole type to the eight-way variety. Though the shell is common to all, the layouts (Figure 6.1), are not. The original types are the three-way and the 180° five-way connectors, which had the merit of allowing a three-way plug to be inserted into a five-way socket. Later types, however, have used 240° pin configurations for five, six and seven connectors, four-way and earth types with the pins in square format, and a five-way domino type with a central pin, along with the eight-way type which is configured like the seven-way 240° type with a central pin added.

This has detracted from the original simplicity of the scheme, which

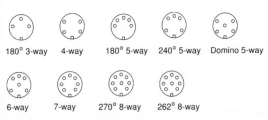

Figure 6.1   DIN connectors.

180° 3-way   4-way   180° 5-way   240° 5-way   Domino 5-way

6-way   7-way   270° 8-way   262° 8-way

115

was intended to make the connections to and from stereo domestic audio equipment easier. The more crowded layouts of DIN plugs and sockets are notoriously difficult to solder unless they have been mechanically well designed, using splayed connectors on the chassis-mounted sockets and to some extent also on the line-mounted plugs. If you frequently need to work on DIN connectors, some form of jig is worth its weight in solder. The five-way 240° type is widely used for audio equipment other than professional-grade equipment, but only where signal strengths are adequate and risk of hum pickup is minimal. Latched connectors can be obtained to avoid the possibility of pulling the connectors apart accidentally. For low-level use, phono plugs are preferable.

Professional audio (or high-quality domestic audio) equipment is more likely to use the XLR series of connectors which provide multiple connections with much superior mechanical quality. These are available as three-, four-, or five-pole types and they feature anchored pins and no loose springs or set screws. The contacts are rated to 15 A for the three-pole design (lower for the others) and they can be used for a maximum working voltage of 120 V. Contact resistance is low, and the connectors are latched to avoid accidental disconnection. There is a corresponding range of loudspeaker connectors to the same high specifications. A variety of other connectors also exists, such as the EPX series of heavy-duty connectors and the MUSA coaxial connectors. These are more specialized, and would be used only on equipment that is intended to match other items using these connectors.

Computer and other digital signal connections have, at least, reached some measure of standardization on the PC type of machines (IBM clones and compatibles) after a long period of chaos. The use of edge connectors is now confined to internal connections because edge connectors are much too fragile for external use, and fractures of board edges were a common problem with older computer designs.

The Centronics connector is used mainly for connecting a computer to a printer, and it consists of a 36-contact connector which uses flat contact faces. At one time, both computer and printer would have used identical fittings, but it is now more common for the 36-pin Centronics socket to be used only at the printer end. At the computer, a 25-pin subminiature D-connection is used, usually with the socket chassis mounted. In a normal connection from computer to printer, only 18 of the pins are used for signals (including ground). The shape of the body shell makes the connector irreversible.

The same 25-pin D-connector can be used for serial connections, but more modern machines use nine-pin subminiature D-sockets for this purpose, since no more than nine pin connections are ever needed nowadays. For other connections, such as to keyboards, mice and monitors, DIN-style connectors, with all their accompanying problems, are often used, though the subminiature D-type connectors are also common. The D-type connectors are widely available in a range of sizes and with a large range of accessories in the form of casings, adapters and tools, so that their use for all forms of digital signals is strongly recommended.

There are now standard DIN fittings for edge connectors, including the more satisfactory indirect edge connectors that have now superseded the older direct style. The indirect connectors are mounted on the board and soldered to the PCB leads, avoiding making rubbing contacts with the board itself.

Lastly, multiway connectors for computing use often use the IDC type of connector – the letters mean Insulation Displacement Connector. These are supplied in two halves, with one half containing the metal connector pins and the small V-grooves for each wire of the multiway cable (Figure 6.2). The cable is laid over these grooves, taking care to ensure that the marked end of the cable is located on the connector for pin 1. The cable should be pressed into place by hand to check that each wire is in its correct position, and then the other half of the shell is fitted on. The two halves are then clamped tightly together until they lock. During this action, the edges of the grooves penetrate the insulation of each wire in the cable, making the connection. A specialized clamp is best for making up connectors, but a Mole wrench can be used if you need only a few connectors made up.

Careless handling of IDC connectors can result in open circuits, and the only remedy is to buy or make up a new cable, because attempting to solder the connections is futile. The main problems with IDC connectors are caused by failure to locate the cable precisely in the holder, and if a large number of cables are to be made up, a combined jig and clamp tool is necessary. Sometimes imperfect cable insulation will cause shorting or an open circuit, and each cable should be tested for continuity and for shorts once a connector has been put on each end.

TUNERS

Basic requirements

The quality of a tuner for a Hi-Fi system is judged on issues of selectivity, sensitivity, stability, adjustability, clarity, noise and linearity. Selectivity is the ability to select the wanted signal from all others, and sensitivity is required to be able to obtain an adequate audio signal from a small RF input. Stability is needed to ensure that the wanted signal remains in tune for as long as required, and adjustability ensures that the tuning can easily be adjusted to find any transmission in its frequency range. Clarity means that the reception will be free of interference, and the tuner circuits should also be free of noise. Linearity is required so that the signal is not distorted at any stage in the tuner action.

The performance of any tuner is critically dependent on the input

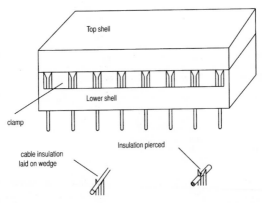

Figure 6.2   IDC connector.

signal level, and for any position that is not within a few miles of the transmitter, the signal strength becomes progressively weaker as the height above ground level of the receiving aerial is reduced, so that the higher the receiver aerial can be situated the stronger the signal input and the better the performance of the tuner. Before tackling any complaints on tuner performance, therefore, check the signal level from the aerial. Now that so many households are using loft-mounted distribution amplifiers, check that the FM signal is not being routed through an amplifier intended only for UHF use.

Except for a small area close to a transmitter, the signal that reaches a receiving aerial consists of a mixture of direct waves travelling in a straight line from transmitting aerial to receiving aerial, and waves that travel from the transmitting aerial to the ionosphere and are then reflected to the receiving aerial. Because the ionosphere is not at a fixed height, the phase of the reflected wave is continually varying, and this in turn means that the incoming signal will always fluctuate in strength unless the receiver is very close to the transmitter.

The classification of radio waves by frequency, (Table 6.1) is approximate, but it corresponds to the designations that are used by broadcasters. The table also shows the designation and approximate height of the atmospheric layers that reflect these waves most strongly. Table 6.2 shows the existing allocations of frequencies for commercial broadcasting.

Tuners used in Hi-Fi systems will operate with VHF signals (the FM bands), and the provision of LF and MF bands is dying out because of the poor reception conditions in these bands. VHF signals, apart from very short range (line of sight) paths, are very heavily attenuated due to signal-absorbing objects such as trees, houses, and hills. Transmitters will use the same frequency only if they are located at sites which are distant from each other, though very low-power repeaters may use a frequency which is shared by a main transmitter which is less remote. There are now some 450 radio transmitters in the UK alone, using LW, MW and Band 2 VHF frequencies, and this number is steadily increasing. Complaints of poor reception can sometimes be dealt with immediately by selecting another frequency for the same transmission.

These services could not be contained within the wavelength allocations on the LW and MW bands for the UK, and it was this, rather than the superior quality of signal that was possible, that forced the change to VHF band and the use of FM. Even VHF/FM radio can suffer from interference from distant transmitters, due to periods of

Table 6.1 Frequency designations and reflecting layers

Designation	Range	Reflecting layer(s)	Height range (km)
VLF	3–30 kHz	D	50–90
LF	30–300 kHz	D	50–90
MF	300–3000 kHz	E, F	110–400
HF	3–30 MHz	F	175–400
VHF	30–300 MHz	None	Note: reflection in peak sunspot times
UHF	300–3000 MHz	None	Note: reflection in peak sunspot times
SHF	3–30 Ghz	None	

Table 6.2 Frequency allocations for UK commercial broadcasting

Designation	Frequencies	Band name
LW	150–285 kHz	
MW	525–1605 kHz	
SW	5.95–6.2 MHz	49 metre band
SW	7.1–7.3 MHz	40 metre band
SW	9.5–9.775 MHz	30 metre band
SW	11.7–11.975 MHz	25 metre band
SW	15.1–15.45 MHz	19 metre band
SW	17.7–17.9 MHz	16 metre band
SW	21.45–21.75 MHz	13 metre band
SW	25.5 –26.1 MHz	11 metre band
TV	41–68 MHz	Band I
FM	87.5–108 MHz	Band II
TV	174–223 MHz	Band III
TV	470–585 MHz	Band IV
TV	610–960 MHz	Band V
TV	11.7–12.5 GHz	Band VI

intense ionization band in the lower parts of the ionosphere, known as 'sporadic E'. This effect is caused by sunspot activity and is to some extent predictable. These interference effects can spark off a number of service requests, so that service engineers should be aware of the problem, which cannot be attributed to any faults in a tuner.

Using FM

The problem of congestion, plus the attraction of being able to broadcast with a wider frequency range, has forced a move to VHF/FM for a large number of transmissions. The internationally agreed standards for FM broadcasting are:

- FM frequency deviation is 75 kHz for 100% modulation.
- The stereo signal is encoded using a 38 kHz subcarrier system, with a 19 kHz ± 2 Hz pilot tone whose phase stability is better than 3° with reference to the 38 kHz subcarrier.
- Less than 1% residual 38 kHz subcarrier signal may be present in the composite stereo output.

FM broadcasts in Europe specify a 50 µs transmission pre-emphasis (the USA and Japan standards use a pre-emphasis time constant of 75 µs, so that personally imported equipment will need adjustment.). The permitted transmitter bandwidth is 240 kHz, and this allows an audio bandwidth on a stereo signal of 30 Hz-15 kHz at 90% modulation levels. At lower modulation levels the high-frequency limit can be extended to 18.5 kHz – this latter limit is imposed by the use of the 19 kHz pilot tone in the stereo coding system.

TECHNIQUES

Selectivity

Ideally, a tuner would have a constant high sensitivity at the operating frequency, along with enough bandwidth on either side of this for the sidebands, but with zero sensitivity at all frequencies outside this range. The tuner techniques that can be used to achieve this (unattainable) ideal include:

- the inductor–capacitor (*LC*) parallel tuned circuit
- transversal SAW filters
- resonant SAW filters

of which the transversal SAW type is now the most common. This type of filter provides an excellent bandpass response, and is well suited to use with ICs.

The basic tuner circuit is invariably a superhet, the circuit devised in 1918 by Edwin Armstrong who also invented frequency modulation. The snag of the superhet system is that it uses a non-linear mixing stage, at whose output four frequencies are present. These are:

- The incoming modulated signal.
- The unmodulated oscillator frequency.
- The modulated sum frequency.
- The modulated difference frequency.

The difference frequency is the desired IF, and though the other frequencies can be easily filtered out, an image frequency is not so easy to deal with. An image frequency is an incoming frequency which will generate the same IF as the wanted frequency. For example, taking for simplicity an IF of 10 MHz, a wanted 100 MHz signal can be mixed with a local oscillator signal of 110 MHz to form the 10 MHz IF, but an incoming signal of 120 MHz will also generate the same IF, and this 120 MHz signal is the image.

The selectivity of any tuned circuits ahead of the mixer stage must be good enough to reject these image (or second channel) frequencies. One aid to image rejection is the choice of a fairly high IF frequency, hence the choice of 10.7 MHz for VHF/FM, which places image frequencies mainly outside the range of the FM broadcast band. Always check for the possible presence of a local image frequency when a tuner is giving interference problems.

Sensitivity

The most important features that affect the sensitivity of a tuner are:

- The S/N ratio.
- The gain for the stages preceding the demodulator.
- Intermodulation effects.

A good S/N ratio will prevent small incoming signals from being swamped by the noise in the mixer stage or any premixer stage. High gain in the stages ahead of the demodulator will ensure optimum conditions for the demodulator, unaffected by change in signal amplitude. Low intermodulation will ensure that the wanted signal is not affected by strong signals on adjacent channels.

The level of aerial noise is the ultimate limitation on sensitivity, and a better aerial can usually work wonders, because a more directional aerial can help to reduce man-made interference which originates in a direction that is not the same as the signal direction.

The frequency changer (mixer) stage in a superhet is responsible for a high amount of noise, and intermodulation is likely also if the incoming signals exceed the optimum level for that mixer. Mixer noise was a major problem for older tuners using valves, but semiconductor devices are very much better in this respect. The best performance is obtained from diode-ring balanced modulator circuits using hot-carrier or Schottky diodes as illustrated in Figure 6.3. These combine excellent noise characteristics with the best possible overload margin, but this system is currently used only on professional equipment.

Tuners that are intended for domestic Hi-Fi at VHF frequencies commonly use junction FETs despite their large inter-electrode capacitances. High-quality equipment normally uses dual-gate MOSFETs because their form of construction allows very good input/output screening. Against this, the noise figure and other characteristics of these MOSFETs are not quite so good as those of junction FETs. A few designs make use of IC-balanced modulator systems.

Stability

A superhet system uses a set of selective fixed-frequency amplifier stages, so the local oscillator is the source of any problems of frequency. In particular, any hum modulated on to the oscillator frequency will be present in the output of the tuner, and harmonics present in the local oscillator output can beat with incoming signals to interfere with the wanted signal.

The stability of a conventional oscillator circuit at frequencies above 30 MHz is poor, and an ideal circuit would have the stability of a quartz crystal oscillator, while still allowing frequency variability. This is possible using variations on the phase locked loop (PLL) principle, for which a typical block diagram is shown in Figure 6.4. The phase of an input signal is compared with the output from a voltage-controlled oscillator (VCO). The output from the phase comparator will consist of the sum and difference frequencies of the input and VCO signals. If the difference frequency is low enough to pass the low-pass loop filter, the resultant control voltage that is applied to the VCO will pull it into synchronous frequency in quadrature phase with the incoming signal provided that the loop gain is high enough. In this condition, the loop is *locked*.

This circuit generates a local signal at the same frequency as an incoming signal, but with much larger amplitude. It can also generate an oscillator control voltage which will accurately follow any variations in the input signal frequency when the loop is locked, and this

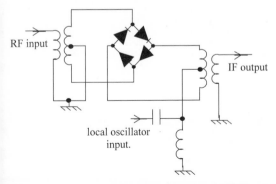

Figure 6.3 A diode-ring modulator used on professional tuner equipment.

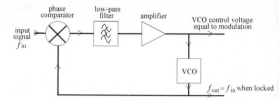

Figure 6.4 PLL type of oscillator circuit.

provides an excellent method of demodulating an FM signal. The application to a local oscillator is illustrated in Figure 6.5, showing a frequency divider stage placed between the VCO and the phase comparator.

When the loop is locked, the VCO output frequency will be a *multiple* of the incoming frequency. A PLL frequency synthesizer uses a crystal-controlled oscillator as the reference, feeding the phase detector through another frequency divider. In this circuit which uses division by factors n and m, when the loop is locked, the output frequency will be $(n/m) \times F_{ref}$. If the division ratios n and m are large enough, the VCO output can be held to the chosen frequency, with the stability of crystal control and with any degree of precision required. This type of frequency synthesizer circuitry is available in single IC form so that this type of frequency control is now commonplace in high quality FM tuners.

Because of the switching actions inside the IC synthesizer which generate harmonics, synthesizer systems can be responsible for producing tuning whistles, and very good screening of the synthesizer chip is needed to keep it down to an unobtrusive level. If any screening is disturbed in the course of servicing, it should be carefully restored.

Frequency drift is less of a problem for FM tuners than it would be for AM tuners working at the same frequency because of the larger bandwidth of the FM stages, though distortion of the received signals will become noticeable if the set is not correctly tuned, or if it drifts off tune during use. Cheaper tuners which do not use the PLL type of frequency synthesizer will have varicap diodes that control the oscillator frequency. These diodes have characteristics that are strongly temperature dependent, so that thermally compensated DC voltage sources are needed for the tuning. Drift problems on such tuners can often be traced to failure of the thermal compensation or use of the tuner in a hot place (such as directly above the power amplifier).

Another problem arises from the time constants that are used in the tuner's AGC system. Ideally, the integrating time constant of the system should be as short as practicable. If the response is too rapid,

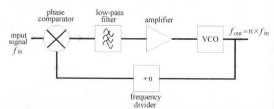

Figure 6.5 Using a PLL local oscillator stage.

however, a low-frequency modulation of the carrier can operate the AGC, so that the lower audio notes are attenuated. Designers attempt to reach a compromise between the speed of the AGC response, and the lowest anticipated modulation frequency which the tuner is expected to reproduce, usually 30 Hz for good quality tuners.

AFC

AFC is widely used, especially in the case of tuners that use varicap diodes in the local oscillator circuits. It does, however, have snags. Using a DC bias on semiconductors alters their characteristics, so that the AFC system can cause a loss of bass response. In addition, the AFC action may try to tune the oscillator to an incorrect frequency because the demodulator is not correctly aligned, and the demodulator is not exactly an obvious place to look when the fault in a tuner is mistuning. This problem is a common one and is just another reason for the use of better demodulator circuits in the best tuners.

Adjustability

The low cost of IC frequency counter systems makes it possible for even the cheaper tuners to incorporate a frequency-meter display. A typical circuit is illustrated in Figure 6.6. This uses a crystal-controlled oscillator, operating usually at 32.768 kHz (the standard frequency for quartz watches), to generate a precise time interval, such as one second. The signal from the oscillator of the superhet receiver is clocked over this period by a counter circuit, and the IF frequency subtracted (since the oscillator frequency is nearly always higher than the incoming signal frequency) so that the counter shows the incoming signal frequency on an LCD or LED display. Latching is used to prevent small changes in frequency from affecting the display, and to ensure that the display is updated at intervals.

The same display can also be used to show information such as the band setting of the receiver, whether it is on AM or FM, whether mono or stereo decoding has been selected, or the local clock time. It is fairly simple to extract time information from the Rugby VLF transmitter so that such a display will always show precise time, automatically adjusting to the change to and from summer time – such circuits are even incorporated into some low-cost alarm clocks.

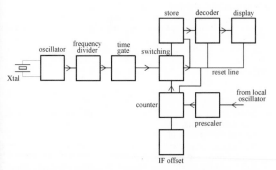

Figure 6.6 Typical tuning frequency display system.

Clarity

Compared to AM radios, an FM radio has much better rejection of external RF noise such as is caused by older car ignition systems or lightning. Rejection of impulse noise depends on the fact that such noise is amplitude modulated with virtually no frequency variation. FM receivers are designed with amplitude limiter stages and a demodulator circuit that is insensitive to AM, providing an AM rejection ratio of at least 60 dB (1000:1) for a good design. This also makes it unnecessary to use AGC unless the designers are attempting to cover every eventuality.

The capture effect is a further benefit of a well-designed FM receiver. Weaker FM signals on the same channel frequency as the desired signal will be completely ignored by the tuner. This effect is measured as the 'capture ratio', expressed as the minimum voltage dB between a more powerful and a less powerful FM transmission on the same frequency which will allow the weaker transmission to be ignored. The capture ratio figure depends on the phase and amplitude linearity of the RF and IF stages prior to limiting, and also on the type of demodulator that is used. A figure of 2 dB is acceptable, and an exceptional design can provide 1 dB. Low-cost tuners may be capable of 3–4 dB or worse. The capture ratio can be degraded if there is overloading in the IF stages.

Linearity

Since the audio output from an FM tuner is likely to be fed to an amplifier and loudspeaker of good quality, the linearity of the FM tuner is important. Good linearity depends on a large number of factors, many of which are concerned, as you might expect, with the demodulator. There are, however, other contributing factors.

The stability and phase linearity of the RF, IF and AF stages must be good. If the FM tuner uses a phase-sensitive demodulator, any RF or IF instability will cause severe phase distortion as the signal passes through the frequency at which there is a risk of oscillation. If junction FETs are employed rather than dual-gate MOSFETs, the design will have to incorporate some method, such as a small inductance in the source lead, to neutralize the residual drain-gate capacitance. Any disturbance of this component will alter the characteristics adversely.

The most important single cause of non-linearity, however, is likely to be demodulator design. There are many types of demodulator known, but only a few are likely to be used in FM receivers that have been manufactured commercially in the last 20 years.

The Foster-Seeley demodulator, illustrated in Figure 6.7, was devised so as to provide more AM rejection by making its output depend on the phase changes induced in a tuned circuit by alterations in the frequency of the incoming IF signal. In this type of circuit the tuned circuit L3C1 provides a balanced drive to the matched pair of diode rectifiers (D1 and D2). These diodes are arranged in opposition so that any normal AM effects will cancel out. A subsidiary coil, L2, feeds the centre tap of L3, so that the induced signal in L2, which will vary in phase as frequency changes, will either reinforce or reduce the voltages that are induced in each half of L3, by altering the position of the electrical centre tap.

The ratio detector circuit, illustrated in Figure 6.8, looks very similar to the Foster-Seeley arrangement. In the ratio detector, however, the diodes are connected so that they produce an output voltage which is opposed, and balanced across the load. The output response is very similar to that of the Foster-Seeley circuit, but it has a greatly improved

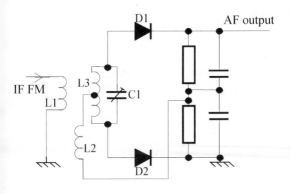

Figure 6.7 Foster-Seeley demodulator circuit.

AM rejection. The ratio detector has been for many years the standard type of demodulator circuit for FM receivers, and it offers a very good internal noise figure.

The phase locked loop demodulator (PLL) uses a circuit such as was illustrated in Figure 6.4. If the voltage-controlled oscillator (VCO) that is used in the loop has a linear *input voltage output frequency* characteristic, then when this is locked to the incoming signal the control voltage that is applied to the VCO will be an accurate replica of the frequency changes of the input signal, within the limits imposed by the low-pass loop filter. This signal is therefore the demodulated audio signal.

The great advantage over all of the other demodulator systems is that the PLL demodulator is sensitive only to the instantaneous input *frequency*, and not to the input signal *phase*. This leads to a very low figure for THD, and it greatly reduces the cost, for a given perform-ance standard, of the FM receiver system. Such a circuit also has a very high figure for AM rejection and capture ratio, even when off-tune, provided that the VCO is still in lock.

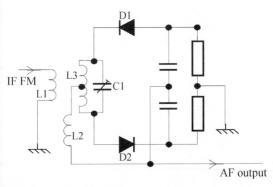

Figure 6.8 Ratio detector.

Pulse-counting demodulators

Pulse counting is an old technique which has been revived, making use of modern digital ICs. A typical circuit layout, due to Pioneer, is shown in Figure 6.9. The 10.7 MHz IF signal is frequency doubled, so as to double the modulation width, and then mixed down to 1.2 MHz using a stable crystal-controlled oscillator. After filtering, the signal is cleaned up and converted into a series of short duration pulses, which have constant width and amplitude. The average value of these pulses is the desired composite (L + R) stereo signal. A conventional PLL arrangement can then be used to reconstruct the 38 kHz subcarrier signal, from which the L and R stereo outputs can be derived by the usual type of matrix circuit.

Tuner performance

A good modern FM tuner can provide a THD performance of around 0.1% (mono) and 0.2–0.3% (stereo), with capture ratios of around 1–2.5 dB and stereo separations in the range of 30–50 dB. Failure to achieve these standards, given a good quality of demodulator, points to problems with the alignment of RF and IF stages, or to the use of poor quality components (particularly if components have been substituted in previous servicing) in the RF and IF stages. The typical AF bandwidth should be in the range 20–40 Hz to 14.5 kHz (between −3 dB points). The LF frequency response depends on the demodulator type, and the lowest limits can be reached using PLL or pulse-counting systems. The upper frequency limit is set by the pilot tone frequency rather than by the design of the tuner.

The signal-to-noise ratio is likely to be around 60 dB, though with good design and ideal reception conditions it could be as high as 70 dB. This figure is greatly influenced by the signal strength at the aerial, and the difference between tuners of widely different price levels may not be so noticeable with a better aerial installation.

PREAMPLIFIER

Requirements

Input switching, and the matching of signal levels from different units, require the use of a preamplifier unless all the equipment is from one manufacturer (like the popular Mini and Midi units) or integrated into one unit (a music centre type of design). The preamplifier can be part of the main amplifier or, more commonly, a free-standing and separately powered unit.

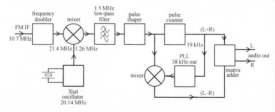

Figure 6.9   Pulse-counting FM demodulator system.

There are no strictly fixed conventions about signal levels and impedances for audio equipment except for professional studio equipment.
Tuners and cassette recorders often use the German DIN standard of
output level, in which the unit is designed as a current source which
will provide an output voltage of 1 mV for each 1000 ohms of load
impedance. An alternative is the line output standard, designed to drive
a load of 600 ohms or greater, at a mean signal level of 0.775 V RMS,
often referred to in tape recorder terminology as 0 VU.

Units that employ DIN-type interconnections will usually (but not
inevitably) conform to the DIN signal and impedance level convention, and units that have phono connectors will not. The usual phono
connector levels are for a minimum load impedance in the range
600Ω to 10k Ω, and mean output signal level in the range 0.25–1 V
RMS. The main exception, as far as domestic audio equipment is
concerned, is the CD player which normally provides an output level
of 2 V RMS. These signal levels should be borne in mind when equipment is to be tested. Complaints of distorted CD sound are usually
due to the user connecting the CD player output to an input that is
designed for much lower signal levels (even a moving coil pickup input)
and causing overload.

Vinyl disc pickup inputs

Though the use of vinyl discs is dying out, discs are still being
manufactured, as are disc players, and there will be a servicing requirement for many years to come. The three main types of pickup cartridge
in use are the ceramic, the moving magnet (or variable reluctance),
and the moving coil. Each of these has different output characteristics
and load requirements. The ceramic type is used only in low-cost equipment.

Ceramic cartridges have a relatively high output signal level, of the
order of 100–200 mV at 1 kHz, and the cartridges are manufactured
with some frequency compensation systems (usually mechanical) so
that they provide a fairly flat frequency response. There is, however,
an unavoidable loss of HF response above 2 kHz, when the cartridge
is connected to a preamplifier input load of 47 kΩ.

The moving magnet or variable reluctance cartridges are designed to
operate into a 47 kΩ load resistance in parallel with some 200–500 pF of
lead capacitance. The actual capacitance of the connecting leads is more
likely to be around 50–100 pF, so additional input capacitance is usually
connected across the phono input socket. This capacitance assists in
reducing unwanted radio signal break-through. Poor response along with
breakthrough points to a failure of this capacitor. The output levels
produced by such pickup cartridges are typically around 3–10 mV at
1 kHz.

Moving coil pickup cartridges became popular for high-quality units
in the last years of the vinyl disc because of their better transient
characteristics and dynamic range. Such cartridges will generally be
designed to operate with a 47 kΩ load impedance. Typical signal
output levels from these cartridges are less than a tenth of the output
from a moving magnet type, so that a very low-noise preamplifier
input, or a separate head amplifier circuit, has to be used to increase
the signal voltage level so that it can be handled by the preamplifier
circuitry.

Inputs

Input to the preamplifier from a CD unit will need only amplification
and impedance level change, but the signals from a vinyl disc player

Figure 6.10   Some simple passive vinyl disc equalizing circuits.

will also need equalization, to the RIAA standard, of the frequency response from a moving magnet, moving coil or variable reluctance pickup cartridge, and tape inputs also need special treatment.

The simplest disc equalizing circuits are passive types, as shown in Figure 6.10(a) and (b). The source impedance must be very low and the load impedance very high compared to R. Circuits which use shunt feedback are illustrated in Figure 6.11(a) and (b). These can accept a wider range of source and load impedances, but have higher noise levels than the passive circuit because of the input load resistor value (usually 47 kΩ). Figure 6.12(a) and (b) illustrate a better way of using negative feedback, using the series circuit so that the input impedance presented to the amplifier is that of the pickup coil. This provides a lower thermal noise figure. Another *CR* network is needed at the output of the equalization circuit (shown dotted) to ensure precise RIAA response.

Other circuits divide the equalization circuit into two parts, using one to shape the 20 Hz–1 kHz section of the response curve followed by a circuit which will shape the part of the response curve that lies between 1 kHz and 20 kHz. Another approach to preamplifier design is to use a very low-noise IC op amp as a flat frequency response input buffer stage. This amplifies the input signal to well above noise level, so that simpler methods, uncomplicated by impedance effects, can be used for equalization.

Audio designers at one time used application specific audio ICs (ASICs) to reduce the cost of RIAA stages and other circuit actions. ASICs are no longer used in the higher quality designs because manufacturers had a habit of discontinuing the supply of ASICs when sales dropped, or would replace them by other ICs which would not necessarily be either pin or circuit function compatible. Such ASIC circuits are a major headache for servicing, and make some circuits virtually unserviceable if the ASIC fails. Modern units are likely to use op amps such as the Texas Instruments TL071 and TL072 which

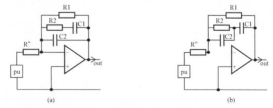

Figure 6.11   Equalizing circuits using shunt feedback.

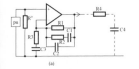
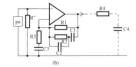

Figure 6.12 Series feedback equalizing circuit.

are of a standardized layout. Such op amps are constantly being improved in performance, but remain compatible both physically and electrically.

CIRCUIT METHODS

Circuits for low-cost equipment are based on ICs, and are fairly standardized, so that servicing amounts mainly to replacing passive or active components as required. For high-quality equipment, however, discrete devices are used to a much greater extent, and are likely to be found when servicing is required. Every designer is likely to work to a different set of compromises, so that the variations between circuitry of 'equivalent' units can be quite remarkable. There is no substitute for good service manuals for such equipment, but some design points can be noted in this chapter for guidance purposes.

The input circuits for moving coil pickups present problems for designers, and the four main methods that are used are:

- Step-up transformer.
- Parallel input transistors.
- IC super-match inputs.
- Very low-noise IC op amps.

There are in addition some very ingenious 'one-off' designs.

Transformers are noiseless, and their output voltage can be well above the level needed to swamp the noise of a preamplifier stage. The main disadvantages are

- Picking up mains hum even when the transformer is well shrouded.
- Poor response to transients.

Good design is also required to avoid the non-linear magnetic effects of the core particularly at low-signal levels. Ortofon have used this type of input since introducing moving coil pickups.

Circuits which use matched-paralleled input transistors (to reduce the base-emitter circuit resistance) have been used by Ortofon, Linn/ Naim and Braithwaite, and an example is shown in Figure 6.13. Care must be taken over matching if one transistor fails and has to be replaced. If matching is not used, the circuit must be devised so that the variation of base-emitter voltage has little or no effect on the currents. This can be achieved by using individual collector-base bias current networks (Ortofon) or individually adjusted emitter-resistors (Linn/Naim and Braithwaite). Again, these expedients call for care if resistors or transistors have to be replaced.

Super-matched input ICs use a set of transistors formed on the same chip and having virtually identical characteristics. Two such transistors can be paralleled to give a very low-impedance, low-noise,

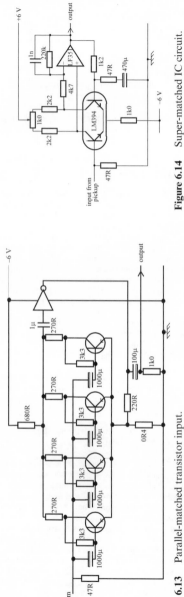

Figure 6.14 Super-matched IC circuit.

Figure 6.13 Parallel-matched transistor input.

130

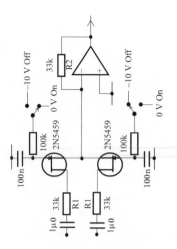

Figure 6.17 FET switching system.

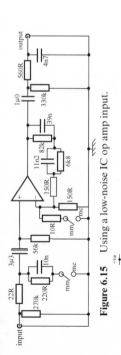

Figure 6.15 Using a low-noise IC op amp input.

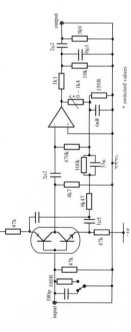

Figure 6.16 Quad 44 low-noise input.

131

matched pair. This method is used in the National Semiconductors LM 194/394 super-match pair, and a typical circuit is illustrated in Figure 6.14. This IC offers excellent input noise performance.

Very low-noise IC op amps, such as the Signetics NE-5532/5534, the NS LM833, the PMI SSM2134/2139, and the TI TL051/052, are intended specifically for audio circuits. A low value of input load resistor is used, and the gain of the RIAA stage is increased in comparison with that needed for higher output pickup. A typical example is illustrated in Figure 6.15.

An unusual input circuit is used in the Quad 44 for ultra-low-noise input. This is based on the fact that bipolar junction transistors will operate satisfactorily at low-input signal levels, with their base and collector junctions at the same DC potential. The type of circuit that is used is shown in Figure 6.16.

Switching the inputs

The traditional method of switching in a preamplifier was a rotary switch, but it is difficult to achieve good isolation between switching units. Some manufacturers have settled for separate switches that are mechanically interlinked, but a much more common system now is the use of FET switches, and a typical arrangement is shown in Figure 6.17. If resistors R1 and R2 are of high enough values, they will swamp the non-linearity of the FET channel and the harmonic and other distortions introduced by the switch will be typically less than 0.02% at 1 V RMS and 1 kHz.

Quietness of operation is an essential requirement for all switching, so that care needs to be taken to ensure that all of the switched inputs are at the same DC level, preferably at earth level. This leads to the use of capacitor inputs on all lines, and any leakage in a capacitor will result in switching transients on that channel.

AMPLIFIER

The main requirements of a voltage amplifier stage are:

- linearity,
- freedom from overloading,
- a good frequency response,
- the required amount of voltage gain.

A simple basic circuit is illustrated in Figure 6.18, using an input transistor, Trl, in common-emitter mode to supply a small voltage signal to the base of Tr2, a common-collector stage that uses a PNP

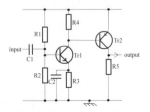

Figure 6.18 A much-favoured transistor pair design for voltage amplification.

transistor. The small voltage swing makes the signal distortion due to Tr1 low, and Tr2, the main voltage amplifying stage, is driven from a relatively high source impedance. Feedback is usually added to this basic circuit to stabilize the working conditions and the signal gain. An improved version of this simple circuit uses a long-tailed pair in place of the first transistor.

More elaborate designs, such as some based on the use of 'current-mirror' circuits, have been used, but since current-mirror circuits are used extensively in IC op amps it seems pointless to have to assemble discrete versions, and such circuits are a headache for the service engineer unless very clear guidance is available in a service sheet.

Designers of high-quality audio equipment were once reluctant to use ICs in audio circuits. This is in spite of the fact that a signal obtained from a tuner or from a vinyl disc might have passed through a large number of crude op amp stages in the studios. Older op amps certainly had a high noise level compared to discrete transistor circuits, but more recent designs are substantially improved. The standard of performance of recent IC op amps is comparable with the best discrete component designs, especially at relatively low closed loop gain levels, so that ICs can now be found in the highest-quality audio circuitry. This makes for much easier servicing.

OUTPUT STAGES

The standard complementary and quasi-complementary single-ended push-pull stages that are used for power output have been illustrated previously in Chapter 3. In many ways the quasi-circuit is preferable to the use of fully complementary output transistors because NPN and PNP power transistors are not truly equivalent, and there are significant differences in the HF performance of supposedly equivalent transistors. Power MOSFETs have a much better maximum operating frequency and linearity than bipolar junction transistors, and they are consequently widely used in Hi-Fi output stages.

The two common forms of construction are the vertical or 'V' MOS-FET, which can employ either a 'V' or a 'U' groove shape, or the 'D' MOSFET. These are typically the 'enhancement' type, in which no current flows at zero gate/source bias voltage, but which begin to conduct as the forward gate voltage is increased. Once conducting, the voltage/drain current graph is very linear. Such MOSFETs can be made as P-channel or N-channel types, allowing true complementary circuits to be constructed, and have the further advantage of showing no stored-charge effects. This, in turn, permits greater speed and lower internal phase and so allows greater amounts of negative feedback to be used, with consequent improvement in performance. The main hazard is parasitic oscillation which can cause rapid failure.

OUTPUT STAGE PROTECTION

For all junction transistors, forward voltage of the PN junction decreases as its temperature is increased. If the power dissipation is significant it causes heating of the base-emitter junction, the forward voltage drop will decrease and current through the junction will increase. Inevitably, some parts of the junction will heat more than others so that current flows preferentially through these parts causing further selective heating, the start of thermal runaway and secondary breakdown. The permitted regions of operation for any particular bipolar transistor type will be specified by the manufacturers in a 'safe operating area' (SOA) curve, of the type shown in Figure 6.19

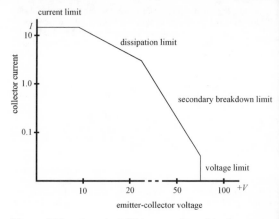

Figure 6.19 A typical SOA curve for a bipolar power transistor.

Circuits which protect the output stage by ensuring that these SOA limits are not exceeded are a common feature of modern power amplifiers, and faults in these circuits can cripple the output stage as surely as faults in the amplifying circuits themselves. For some purposes, the use of fuses can offer some protection, but more elaborate systems are needed for high-power outputs, using some form of clamp circuit between base and emitter of the power transistors. Power MOSFETs can use a much simpler protection in the form of Zener diodes as illustrated in Figure 6.20.

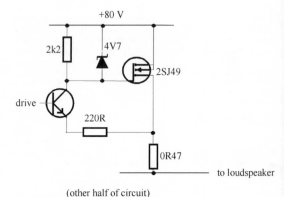

(other half of circuit)

Figure 6.20 Using Zener diode clamps for a power MOSFET circuit. Only one of a complementary pair is shown, and details of surrounding circuitry are omitted.

OTHER TECHNIQUES

Phase shift and slew rate

All amplifiers which use negative feedback are likely to suffer from oscillation if the negative feedback turns to positive because of excessive phase shift at some frequency at which the amplifier still has a gain value more than unity. At one time the usual precaution was to connect a small capacitor between base and collector of one transistor in the feedback loop. This, however, causes more problems than it solves, because the added capacitor must be charged and discharged, and since the rate of charging/discharging is limited by the time constant of this capacitor and the resistance that feeds it, this time constant determines the slew rate of the amplifier.

A slew rate that is too slow will cause a feedback amplifier to lose gain for the duration of a sharp change of signal level, and this effect is painfully obvious to the ear. Phase-correcting capacitors need to be positioned so that they will have the least possible slew rate effect, and the tendency nowadays is to use lower levels of overall negative feedback so that there is less likelihood of oscillation when phase shift becomes severe.

The *RC* Zobel network of Figure 6.21 is also used for correcting the phase characteristics of the feedback loop. This is connected across the output and helps to avoid instability with an open-circuit load, and to this is added the inductor/resistor network in series with the load to avoid instability on a capacitive load. The inductance value is typically 4.7 µH so that it has no effect in the audio range, but if you are using square-wave testing this portion of the circuit should not be included because it will round off the edges of the square waves. A further precaution is the use of a simple low-pass filter at the input to avoid slew rate effects caused by frequencies beyond the audio region.

Power supplies

The normal practice for audio amplifiers in the past has been to use the conventional power supply circuit consisting of a mains transformer, bridge rectifier and reservoir capacitor. The output voltage of this circuit will decrease as the load current increases, and the no-load voltage will therefore be considerably higher than the voltage at maximum rated load. In addition, the AC ripple will increase as the load current increases, and the internal resistance of such a supply will make it possible to cause unwanted signal paths between sections of the amplifier.

The use of a stabilized supply with a foldback characteristic (see Chapter 4) provides a constant voltage level with protection against overload and very low internal resistance, so that this type of circuitry is increasingly used in high-quality power amplifiers.

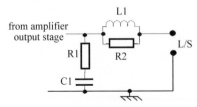

Figure 6.21   The Zobel *RC* circuit and *LR* stabilizing circuit.

Unorthodox circuits

The Class B single-ended push-pull type of circuit is so common that you might assume that nothing else exists. In fact, there are as many variations of output stages found as for input stages, and for some of these servicing is complicated by their unfamiliarity. Class A circuits are, for example, quite often found, particularly for lower power levels, and some high-power types also exist, providing a significant degree of room heating as well as audio power. Of the unorthodox circuits, however, the most intriguing is the Quad type known as the current-dumping amplifier, which dates back to 1975. Experts still argue about the precise way that this output stage operates, but no one denies that it does work very well. The accepted explanation of the system has been provided by Baxandall.

The basic circuit is as shown in Figure 6.22, with a linear amplifier, shown here as an op amp, driving a Class B pair of complementary transistors which are unbiased and connected to a load through a resistor. By choosing the correct values of the resistors in this basic circuit, all distortion due to the lack of bias will vanish. In effect, the linear amplifier supplies the load while the main power transistors are cut off, and the characteristic slope can be made the same as it is when the main transistors are conducting.

The snag of this simple system is that the resistors would dissipate too much power, and the Quad solution is to use an inductor in place of R4 and a capacitor in place of R2. The performance of the Quad 405 amplifier thoroughly proved the value of this type of design, and has spurred other designers to try other methods of combining unbiased power transistors with linear drivers, using a technique called *feed-forward*. In a feed-forward amplifier, the final output is compared to the output of a linear stage, and correcting current is fed in to minimize the difference. Several variations exist and one such amplifier is illustrated in Figure 6.23, due to Technics.

Circuit faults

Total breakdown in an audio amplifier, leading to no sound or a very distorted sound, is relatively easy to diagnose and cure, but faults that lead to increases in distortion that are barely audible are much more difficult to deal with, and require more elaborate testing equipment than the average servicing workshop carries. At one time, the total harmonic distortion (THD) figure was thought to measure the quality of an amplifier, and a figure of 0.1% was deemed acceptable.

Certainly it is useful to be able to measure THD in the workshop, but although a high THD figure indicates problems, the THD alone is not a good guide to how the amplifier sounds. Intermodulation distortion (IMD), on the other hand, has a large effect on the sound clarity,

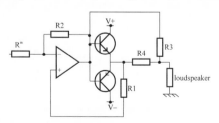

Figure 6.22   Basic current-dumper circuit.

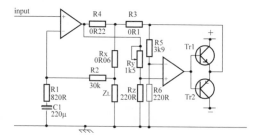

Figure 6.23 A Technics output stage using feed-forward.

and when a user complains of 'muddy sound', this may indicate faults that lead to IMD, caused by non-linear behaviour in the signal path. For a high-quality design, the total IMD and THD should be below 0.01% over the range 30 Hz–20kHz.

Another important problem arises from transient signals. When one stage of the amplifier saturates because of a transient the signal momentarily is interrupted, and the measurable effect is called transient intermodulation distortion or TID. This sounds to the ear as a 'tizz' in the signal. This type of defect is most likely to occur when amplifiers use large amounts of negative feedback along with a stage which is subject to slew rate limiting. Transients can also cause objectionable ringing or overshoot effects, usually when the stability is marginal or when a low-pass filter with a large attenuation rate is used. Careless replacement of transistors with 'equivalents' that have lower slew rates can cause this effect to become objectionable in an amplifier which was previously of good performance.

Capacitors are the weak link among circuit components, both from the point of view of performance and also from the point of view of troubleshooting. The problems arise from manufacturing methods and include the inductance that is present when a capacitor contains a foil winding, piezo-electric effects in ceramic types, stored charge in some plastic (particularly polypropylene) dielectrics, and non-linearity of leakage current in many varieties.

Interference

A large portion of complaints about amplifiers are about interference signals. Of these, the most common varieties are noise and unwanted signals, either from the supply line, or by direct radio pick-up. These can often be difficult to eliminate and they may be peculiar to the positioning of the equipment, so that the amplifier behaves flawlessly when it is tested in the workshop. Mains hum can sometimes be due to bypass capacitors developing a high resistance, and replacement must be specified as low effective series resistance (ESR) types.

One odd problem that sometimes crops up is caused by the loud-speaker acting as a microphone and putting signals into the negative feedback loop of the amplifier. The manufacturer of the equipment must be consulted if this fault arises, because there is no universally applicable cure. This is just one of a number of microphonic effects that can appear due to the vibration of the components, and detection can be difficult when the equipment is being tested in quiet surroundings.

Mains-borne interference takes the form of noise pulses when electrical equipment is switched on or off (lifts are the most common culprits). This is usually caused by radio frequency pick-up problems, and can be cured by attention to the signal and earth line paths, assuming that the causes of interference are adequately suppressed.

LOUDSPEAKER UNITS

Loudspeaker faults are mainly mechanical faults, and are very seldom repairable in the workshop, though some repairs that require only gluing may be possible. When a loudspeaker fails, however, the service engineer should find why, and, if necessary, advise the customer so that further failure can be prevented.

Where active speakers are used, the usual amplifier faults can be found, and the main complication is opening the loudspeaker cabinet to gain access to the amplifier. Very great care should be take when resealing the cabinet, because even small air leaks can affect the performance of the loudspeaker. On a larger scale, units such as the Quad ESL-2 loudspeakers contain protective systems, and a failure of one of these systems will cause the output to be at low level and distorted. Repair is possible if good service sheets are available, but return to factory is needed if the fault is elusive.

The usual cause of failure of a conventional loudspeaker is overload, and the two main effects are burning out the voice coil, or damaging the suspension. A burned out voice coil can be checked easily by a resistance test across the speaker terminals, but checking for a damaged suspension requires the casing to be opened. Note that some manufacturers may not accept the return of a loudspeaker which has been opened, so if there is any possibility of a full or partial refund, the loudspeaker should be returned intact.

Figure 6.24 shows a graph of voice coil temperature versus input power. The temperature limits are determined by a variety of factors such as the former material and the temperature rating of the enamel insulation on the wire. A small amount of self-protection is provided by the positive temperature coefficient of resistance for the wire, typically 0.4% per °C change of temperature.

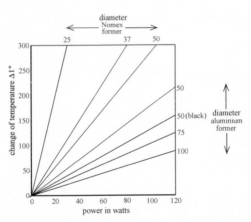

Figure 6.24 Voice coil temperature and input power (typical).

There is no loudspeaker which is totally immune from the effects of hard rock sounds at high dissipation levels, and since many devotees have become deaf in the pursuit of their music they are likely to use ever-increasing volume levels until something gives, whether it is the voice coil, the suspension, or the neighbours. Loudspeaker failure is much less likely when music is played at tolerable levels, and when the user realizes that damage is possible.

CASSETTE SYSTEMS: ANALOGUE AND DIGITAL

Magnetic recording on iron wire has been around since 1898, but was not developed into its present-day form until the 1940s. The improvements hinged on using tape, nowadays a durable form of plastic base which can be coated with magnetic material such as ferric or chromium oxide.

Older open-reel tapes were sold on the basis of tape length, but cassettes (which use a set playing speed) are graded according to the maximum playing time. Table 6.3 shows the total thicknesses of these tapes, whose width is set at 3.81 mm (0.15 inches). These tapes are usually made by slitting down wider tape (up to 48 inches), and unless high-quality slitting machinery has been used, the tape will have uneven width, causing fluctuations in sound level. Always check a faulty cassette machine using high-quality tape, and advise the customer if the tape quality is the cause of the problem.

The three factors that are most important in the tape recording system are the tape speed (which is fixed for cassette tapes), the head gap, and the use of bias. This assumes that the tape is tightly held against the head. Any problems with head gap (caused by wear, for example) or contact of tape with the head (caused, for example, by iron oxide rubbing off the tape on to the head) will show up as a reduced high-frequency range. A worn tape head must be replaced; a dirty tape head can be cleaned, but the cleaning must be done following the manufacturer's instructions, because too enthusiastic cleaning can cause more damage than the dirt. Stories about customers who have cleaned tape heads with kitchen abrasives are probably myths, but the DIY movement has a lot to answer for. Note that if heads are renewed, their azimuth angle must be correctly set up as advised by the manufacturer.

Equalization

Magnetic recording is a non-linear process, and bias is just one of the methods that are used to correct this problem. The ultimate solution is to use digital recording methods, but DAT has been slow to take off in the UK, partly due to the high price of equipment and partly because it has to be added to a system – the old tape system has to be retained in order to play the older recordings and pre-recorded tapes.

Because no tape system can provide a flat frequency response the usual method of equalization is used for all systems of analogue tape recording, quite apart from any noise-reduction circuitry. Tape equalization is nothing like the equalization that is used in FM receivers or for vinyl disc reproduction, where the equalization process used when transmitting or recording is reversed at the receiver or player. The equalization that is used during recording is designed to produce a level of tape magnetization (for signals of different frequency and constant amplitude) which will be fairly even for most of the frequency range, with a bass boost and a fall-off at higher frequencies which depends on the tape speed. The equalization that is used on playing attempts to compensate for the fact that a tape with constant magnetic level will replay so that the higher frequencies have a much greater amplitude than the lower frequencies. The losses of high-frequency signals at the tape head, however, demand an additional boost to the highest range of frequencies, and the overall playback response is typically of the form shown in Figure 6.25.

Equalization circuitry must not be altered unless you are quite certain that it is faulty, and great care needs to be take to use high-grade components and to test carefully afterwards. Test equipment for high-quality cassette recorders is expensive and may need calibration, so that it is not a task to be undertaken lightly.

Bias

The choice of bias frequency is always a compromise on the part of the designer, and nothing should be done during servicing to alter the frequency. Using too low a bias frequency can leave a large enough bias signal remaining on the tape to cause overloading in the replay circuits, and too high a frequency will need a high bias amplitude so as to pass enough current through the inductance of the tape head, causing overheating and saturation. A rough guide is that manufacturers use a frequency which is between four and six times the highest audio frequency that the equipment can handle, and this leads to a frequency of around 120 kHz for high-quality tape/cassette machines.

The shape of the bias waveform is as important as its peak amplitude. Any trace of mains hum on the bias waveform will lead to hum on the tape, and many complaints of hum and distortion arise from bias oscillators which are being modulated or whose waveform is not perfectly symmetrical. One authority has calculated that the amount of noise in the bias waveform should be one fortieth or less of the noise content of the incoming signal; a condition that is not easy to achieve, calling for a very pure sine wave. This quality of sine wave cannot be determined from examining the waveform using an oscilloscope.

Bias level setting

Because the correct setting of bias level is difficult, even if instruments are available, recorder manufacturers provide a switch to select between preset bias values. For a cassette recorder these are labelled as follows:

Type 1 all ferric oxide tapes, used with a 120 μs equalization time constant
Type 2 all chromium dioxide tapes, used with a 70 μs equalization
Type 3 dual coated ferro-chrome tapes, used with 70 μs equalization
Type 4 all varieties of metal tapes

A few machines, particularly the three-head types, contain a test system which uses a built-in dual-frequency oscillator with outputs at 330 Hz and 8 kHz. Recordings can be made using each signal in turn, and the bias set so that the output on replay is the same for

Table 6.3   Tape length and thickness for cassettes

Cassette	Tape length	Tape thickness
C60	92 m	18 μm
C90	133 m	12 μm
C120	184 m	9 μm

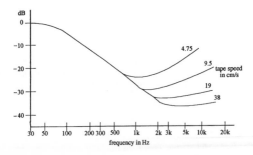

Figure 6.25 Tape replay equalization graphs, with high-frequency boost.

each frequency. On some modern machines this process has been automated, so that each new cassette is tested before a recording is made. Faults in this system can lead to incorrect bias and consequent distortion or muffled sound.

Mechanical faults

Unlike most parts of a modern audio system, the mechanism of a tape recorder plays a large part in enabling high-quality reproduction, just as it did in vinyl disc players. The tape speed must be exactly correct and constant. High-quality tape mechanisms will use a dual-capstan system that has separate speed- or torque-controlled motors. This ensures that the tape maintains a constant tension across the record and replay heads. The other important parts of the mechanism are the drive motors and braking systems for the two reels.

Cassette recorders present the problem that access to the tape is limited by the design of the cassette itself. Dual-capstan drive machines use the erase head access port for the capstan on the rewind side, adding a narrow erase head fitted into the adjacent, unused, small slot in the cassette body.

For a cassette recorder, the pressure pad is an integral part of the cassette itself, so that most designs have to accept this limitation. A few cassette drives, however, use three heads and dual-capstan drive, and the built-in pressure pad is pushed away from the tape when the cassette is inserted. The tape tension is then totally determined by the design of the machine.

On servicing, irrespective of the fault, the chance should be taken to clean the capstan drive shafts and pinch rollers, which are always coated in a film of oxides. At the same time, the heads can be cleaned to ensure the best possible contact between heads and tape.

Typical circuits

The critical electronic parts of any tape recorder design are the replay amplifier, and the bias/erase oscillator. The mid-frequency output of a typical high-quality cassette recorder head will only be of the order of 1 mV, and the S/N introduced by the replay amplifier should be at least 20 dB better than the usual tape S/N ratio of about −52 dB, giving a figure of around −72 dB. This corresponds to a noise level of about 0.25 µV, a figure which is very difficult to meet.

This means that discrete component designs are preferred for high-quality machines, although low-noise ICs may very well replace them in the lifetime of this book. Figure 6.26 illustrates one Pioneer design which has proved suitable for high-quality cassette recorders.

The positioning of the equalization network varies. Some designers place equalization following the input amplifier, others use feedback networks in the input amplifier as a method of equalization. A typical equalization stage using an IC is shown in Figure 6.27.

The bias oscillator must provide a substantial output voltage swing because it will also be used to power the erase head. The symmetry of the waveform must be excellent, and the oscillator must have a very low noise figure. The waveform is always a low-distortion sine wave.

The circuit layout that is most often used is a symmetrical push-pull type, and a typical example is shown in Figure 6.28. Where, as in this example, the highest quality of bias waveform is not quite so important, circuits that use the erase head as their inductor are commonly used. For high-quality equipment, the bias level will be switch-selectable to suit tape types 1–4, along with a fine adjustment to lessen the possibility of recording signal cross-talk from one channel to the other, and a separate inductor will be used to tune the oscillator to the required frequency.

The recording amplifier will almost certainly include a low-pass filter system to prevent higher frequencies interfering with the noise-reduction circuit (see later), but otherwise the design of this amplifier follows the same low-noise principles as the replay head amplifier.

Recording level indicators in modern equipment almost always use instantaneously acting LEDs, with the drive to the indicating circuit adjusted to respond to the peak recording level. This can use any one of several purpose-built ICs. Older equipment used moving coil display meters, but to obtain good performance, the meter movements had to be mechanically matched to the requirements of the system.

Tape speed is always nowadays electronically regulated, using servo systems such as that illustrated in the block diagram of Figure 6.29. The capstan drive motors will either be DC types whose drive current can be increased or decreased, or the brushless AC type. The brushless type is gaining in popularity, and is fed from an electronically generated AC waveform whose frequency can be altered to affect the speed of the motor. The main problem associated with this circuitry is that its complexity makes fault diagnosis and repairs more difficult and costly to perform.

DIGITAL RECORDING

As noted elsewhere in this book, digital encoding is done by sampling the input signal at sufficiently brief intervals, and then representing the instantaneous peak signal amplitude of each sample binary-coded number. The process is often referred to as *quantization*. The quality of the sound that can be obtained depends on the number of samples per second (relative to the highest frequency to be recorded) and the number of binary digits (bits) used to represent the amplitude figures. The research that led to the adoption of the CD standards suggested that 16-bit binary numbers were wholly acceptable, and that for less stringent conditions 13-bit numbers would be adequate. Even this reduced system requires a minimum bandwidth of 448 kHz when a sampling frequency of 32 kHz is used.

The particular advantages of using digital recording with a tape system is that it eliminates all problems of bias and tape non-linearity, eliminates all traces of print-through, and allows recording faults such as drop-outs to be corrected or made inaudible. Add to these points

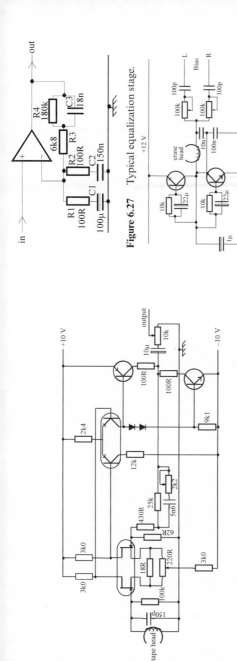

Figure 6.27 Typical equalization stage.

Figure 6.28 A typical bias oscillator circuit.

Figure 6.26 Low-noise cassette input circuit, due to Pioneer.

143

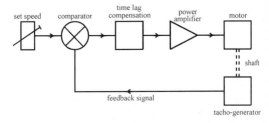

Figure 6.29   Servo system block for tape speed regulation.

the absence of wow, flutter and pitch errors caused by incorrect tape speed and the possibility of processing the signals so that time or phase errors between tracks are eliminated.

A good level of audio recording performance can, incidentally, be achieved in most NICAM stereo video recorders, if the audio signals are entered through the audio input sockets that are usually provided on such machines, or by way of the SCART socket, but paradoxically DAT machines are much more expensive. The progress of DAT was blocked by recording companies who feared that such machines could be used to make perfect copies of CDs, but the fact that video recorders could be used for the same purpose seems to have escaped them.

NOISE REDUCTION IN ANALOGUE SYSTEMS

Modern noise-reduction systems are complementary, meaning that whatever modifications to the signal made before recording or transmitting will be exactly reversed on replay or reception. The audio signal will have passed through both stages, but the noise through only the replay or reception stage. The usual system is referred to as a *compander,* meaning that the audible amplitude range is compressed before recording or transmitting, and expanded when the signal is replayed or received. This system has no effect on noise that is present in the signal before the compression process, and a well-designed system should not degrade the sound in any detectable way.

Companders were originally used in discrete circuit form, but are now invariably ICs, so that the servicing actions consist simply of diagnosing whether or not the chip is faulty, and replacing it if it is. The usual signs of problems are 'breathing' effects, in which sudden changes of noise level can be heard when alternate passages of loud and soft music are played. Modern cassette machines usually offer a choice of Dolby B, Dolby C or Dolby HX noise-reduction systems, all of which should have an excellent performance in the absence of a chip fault. A servicing problem may exist for some older Technics machines which used Dolby with the alternative of the dbx system. If a fault affects the dbx system, it is now difficult to obtain spares, rendering the customer's collection of dbx encoded tapes unplayable.

DIGITAL SOUND BROADCASTING AND SATELLITE RADIO

The standards of quantization that are used for the CD system are designed to provide a very high audio standard, and since the system

144

was developed, much research has been devoted to trying to reduce the bandwidth requirement to allow for digital broadcasting either on existing allocated frequencies or by the way of satellite channels. The NICAM system that is used for TV in the UK is an example of how the bandwidth requirement can be reduced, and the digital cassette systems that have been proposed (some of which are now implemented) are another example. Currently, proposals for digital audio broadcasting (DAB) are being considered.

The existing FM broadcasting system was good at the time it was inaugurated, but the development of CDs has shown up FM transmission as a weak link in the audio chain, and FM communications are more liable to disruption by interference and by ionospheric fluctuations than digital communications. Digital sound can be transmitted along comparatively short paths using conventional transmitters, but will then be subject to the same problems of shadowing as FM transmissions, so that proposals for digital sound transmission almost wholly centre on satellite transmissions.

The use of satellite broadcasting is not a perfect solution, because the normal technique of using a satellite that is rotating at the same angular speed as the earth (a geostationary satellite) provides good reception at low latitudes (near the equator), but poorer conditions at high latitudes, such as in northern Canada. In addition, satellite reception from moving vehicles is much more difficult to achieve, and the choice is to use a tracking aerial or a low-gain omnidirectional type. One proposal is to use a set of four satellites in a very elliptical orbit to avoid the need for tracking aerials; the less acceptable alternative is to reduce the bit rate of transmissions to allow a poorer S/N ratio of transmission at the cost of signal quality.

At present, research in Europe, North America and Japan points to a DAB system that uses a bit-reduction system called MASCAM (masking-pattern adaptive sub-band coding and multiplexing) and a transmission system called COFDM (coded orthogonal frequency division multiplex). The masking effect is a well-known effect of the ear, in which a large amplitude signal drowns out a lower amplitude signal at around the same frequency, and this can be used to reduce the bit-rate by coding only the larger amplitude components in a narrow bandwidth. In a full MASCAM system, the whole audio bandwidth is divided into narrow sub-bands. For each sub-band the spectrum of signal is continually analysed so that audio masking can be predicted, and the bit rate correspondingly reduced. Other such systems are ASPEC (adaptive spectral entropy coding) and MUSICAM (masking-pattern adaptive universal sub-band integrated coding and multiplexing). No decision has been taken at the time of writing, and a compromise system that incorporates the feature of two or more of these systems might be adopted.

The modulation system known as OFMD operates by generating a large number of carriers at equal spacings, each of which can be digitally modulated by audio sub-band codes. One proposed system uses a bit rate of 256 kbit/s for each stereo transmission with a bit error rate of better than 10^{-3}. A DAB service using these methods could be accommodated in the 1.5 GHz or 2.5 GHz bands, and could provide 12 stereo transmissions in a 4 MHz bandwidth, each of higher quality than is currently possible using FM methods. In addition, because COFDM can be transmitted at low power levels and has a spectrum similar to that of noise, terrestrial broadcasting is possible on channels that are also used for other services.

CHAPTER 7

TELEVISION AND TELETEXT RECEIVERS

THE VISION SYSTEMS

For many years now, the world has been serviced by three analogue standard colour television systems with an aspect ratio of only 4:3 and colour gamut equations as standard. However, in spite of the other significant differences, it has been possible to produce a receiver that will, in most cases, function correctly when either NTSC, PAL or SECAM signals are provided as inputs. Minor problems arise with such a multi-standard receiver because a number of substandard variations for all three main systems have been developed to meet local needs.

While the NTSC system is predominant in the Americas, Japan and the Pacific Rim countries, PAL is widely used throughout Europe, much of Africa and Australasia. The SECAM system is chiefly used in France and the French dependencies and Russia. In spite of this broad partitioning, there are many countries that use one of the alternative systems and some are in fact in a process of actually changing systems. While Britain now only employs the UHF Bands IV and V for TV, much of the rest of the world, still use VHF Bands I and III as well. In addition to these differences, it will be recalled that the mains electrical supply voltage and frequencies can also vary. Some of the pertinent signal parameter variations are listed in the Figure 7.1.

Systems H, I, K1, and L all have a vestigial sideband that extends to 1.25MHz whilst for all the others this is reduced to 0.75MHz.

THE SOUND CHANNELS

In a similar way, the sound channels have developed from simple mono transmissions to a range of different stereo or dual language services. Initially, the Zenith GE Pilot tone system used for FM radio broadcasting was adapted to television but had a relatively short life due chiefly to the beat notes produced from the subcarriers. Currently, the German Zweiton analogue system is very popular in PAL regions because the main television sound carrier is modulated by the sum signal $((L + R)/2)$ as a mono input for a standard receiver and the sub-carrier is used only for right (R) channel of the stereo signal. Processing these two components through balancing amplifiers and adders and subtractors then generates the separate L and R signal components. For Systems B and G the second subcarrier is positioned at 5.742 MHz, while for System D it is placed at 6.742 MHz. For System M, the normal sound channel carries the sum signal $(L + R)$ while the difference signal $(L-R)$ is carried on a subcarrier positioned at 4.72 MHz.

In NTSC territories, the Japanese BTSC/ MTS system is popular. This employs an $L + R$ main sound channel, plus a frequency multiplex at subcarrier frequencies based on multiples of the line frequency. The $L-R$ component is amplitude modulated using double sideband suppressed carrier (AMDSB) at twice the line frequency above the main

System	B	D	G, H	I	K, K1	L	M	N
Lines per field	625	625	625	625	625	625	525	625
Frame rate	25	25	25	25	25	25	30	25
Channel bandwidth	7 MHz	8 MHz	8 MHz	8 MHz	8 MHzx	8 MHz	6 MHz	6 MHz
Vision bandwidth	5 MHz	6 MHz	5 MHz	5.5 MHz	6 MHz	6 MHz	4.2 MHz	4.2 MHz
Vision modulation	–ve	–ve	–ve	–ve	–ve	+ve	–ve	–ve
Sound spacing	5.5 MHz	6.5 MHz	5.5 MHz	6 MHz	6.5 MHz	6.5 MHz	4.5 MHz	4.5 MHz
Sound modulation	F.M	F.M	F.M	F.M	F.M	A.M	F.M	F.M

Figure 7.1

sound carrier. A Second Audio Programme (SAP) that represents an audio signal with 12kHz wide audio band which is frequency modulated on to a subcarrier at five times line frequency may also be included.

A further Japanese variant employs the FM–FM system where the subcarrier is locked to the second harmonic of the line timebase. This may carry either the L–R stereo component or a second language in the form of frequency modulation. An amplitude modulated 55 kHz subcarrier is used to convey the necessary control signals to the receiver to manage the automatic switching between mono, stereo or bilingual operations.

The NICAM-728 (Near Instantaneous Companded Audio Multiplex-728) system provides either mono, stereo or dual language audio at near CD quality using digital transmisions. This is achieved with PAL TV systems using a subcarrier 6.552 MHz for System I, and 5.85 MHz for systems B, G and L. In each case, the main sound carrier carries a normal mono signal, while the NICAM subcarrier conveys both Left and Right stereo channels. In all systems, the normal mono sound channel is retained in the interests of receiver compatibilty.

The subcarrier is differentially encoded with the digital signals for both channels of the stereo pair. This sound subcarrier is quadrature (four phase) PSK modulated, where each resting carrier phase represents two bits of data, thus halving the bandwidth requirement. Because the data is differentially encoded (DQPSK), it is only the phase changes that have to be detected at the receiver. The left- and right-hand channels are sampled simultaneously at 32 kHz, coded and quantized separately to 14 bits resolution, and then transmitted in 728-bit frames per millisecond or 728 kbit/s.

THE COLOUR STANDARDS

All three analogue systems use 2:1 interlaced line scanning and a colour signal that is composed of luminance (luma) or brightness (Y), plus two colour difference components (I and Q) forming the chrominance (chroma) or colour information. In all three cases, the luma component provides the monochrome information for any black and white receiver to ensure compatibility.

NTSC system

In this system the image is scanned at a line frequency of 15.734254 kHz and field frequency of 59.94 Hz to produce an image that consists of 525 lines scanned in nominally 16.677 milliseconds. The two weighted colour difference chroma components are referred to as the I and Q signals which are phase shifted relative each other by 90º, these are then used to modulate the subcarrier frequency at 3.579545 MHz using DSBSC. The luma and chroma components are formed using the weighting matrix shown in Figure 7.2.

$$Y = 0.299R' + 0.587G' + 0.114B'$$
$$I = 0.736(R' - Y') - 0.269(B' - Y')$$
$$Q = 0.478(R' - Y') + 0.414(B' - Y') \text{ where R', G', B' and Y'}$$
represent gamma-corrected signals.

Figure 7.2

PAL system

This system is based on 625 line, 50 Hz field scanning rates, again 2:1 interlaced. The line frequency is thus 15.625 kHz. The two U and V weighted colour difference signals are modulated on to the subcarrier of 4.43361875 MHz again using DSBSC, but with the phase of the V component inverted on alternate lines. The Y, U and V components are formed using the gamma-corrected matrix values shown in Figure 7.3.

SECAM system

The Y signal component is again formed using the same values for R', G' and B' but the colour difference components are weighted as follows: $D_R = -1.902(R' - Y')$ and $D_B = -1.505(B' - Y')$ and these are used to frequency modulate two subcarriers at 4.4626 MHz and 4.25 MHz respectively. The numerical values of 1.902 and 1.505 are chosen so that the resulting frequency deviation does not exceed 3.9 to 4.74 MHz for both subcarriers. The minus sign associated with these values indicates that an increasing amplitude colour difference signal causes the subcarrier frequency to fall.

The universal receiver

It will be clear that each of these systems can be automatically identified from the subcarrier frequencies and the methods used for modulation thus allowing for the use of an integrated system processor.

THE TELEVISION RECEIVER

The major aim of the above systems must be to ensure a high level of fidelity for any signal received at a distance over any transmission medium. Thus the design and maintenance of the receiver which is the final element in the reproduction chain is particularly important. It is worth pointing out here that the TV receiver is virtually complete with its own CRO. A careful inspection of the display can provide quite invaluable information about fault conditions. The important requirements for the receiver can therefore be summarized as follows

1. to select the wanted signal from the many others present in any given waveband;
2. to recover the information from the modulated wave,
3. to present the information in a suitable manner, e.g. audio, video, data. etc.,
4. to carry out these functions without degrading the signal-to-noise ratio by more than is minimally necessary.

Since the received noise power is proportional to the system bandwidth, it is important that the receiver bandwidth should not be greater than that required to just accommodate the wanted signal.

$$Y = 0.299Y' + 0.587G' + 0.114B'$$
$$U = 0.493(B' - Y')$$
$$V = 0.877(R' - Y')$$

Figure 7.3

A receiver needs to be responsive to a wide dynamic range of signal amplitudes. The lowest level is set by the thermal noise generated by the receiver input circuits and the simultaneous noise present within the transmission medium due to atmospherics and man-made interference. In today's congested communications bands, these requirements are best met by using the supersonic-heterodyne (superhet) receiver.

It will be recalled that the basic front end of the superhet receiver consists of an RF amplifier, mixer and local oscillator. When the RF and local oscillator signals are combined in the non-linear mixer stage, sum and difference frequency components are generated and either of these can be used as the intermediate frequency (IF) for further processing. While it is completely standard for the lower difference frequency to be selected as the IF in entertainment systems, it is possible that the higher value might well be used in an up-conversion communications receiver. Although the superhet receiver is a standard feature it is not without its problems. Any signal present on the aerial at the intermediate frequency can break through into the IF stages in the form of noise. In the case of a down-conversion receiver, the sum products from the mixer forms an image of the IF on the other side of the oscillator frequency spectrum and can generate image or second channel interference. Furthermore, adjacent RF channel signals can also enter the receiver as noise. The local oscillator frequency can also escape from the receiver as interference to other systems. The RF amplifier stage not only aids the selection of the wanted channel signal, it also helps to minimize local oscillator radiation. By design and international standardization, the IF is chosen to be high enough to give good image rejection properties, as low as possible to accommodate the signal bandwidth and from a little used part of the frequency spectrum, features that aid good selectivity and IF amplifier stability.

The interference rejection ratios are defined by the formula:

$$20 \log(\text{Wanted signal voltage}/\text{Interference signal voltage}) \text{ dB}$$

The RF signal input bands for the various TV servcies are shown in Figure 7.4, and the standardized IFs are shown in Figure 7.5.

VHF bands	Channels 1 to 5	41–68 MHz
VHF bands	Channels 6 to 13	174–216 MHz
UHF band	Channels 20 to 68	470–854 MHz
Cable systems	Band 1	68–88 MHz
" "	Band 2	108–176 MHz
" "	Band 3	230–300 MHz
" "	Hyperband	300–470 MHz

Figure 7.4

Sound/Vision IF	PAL-G	33.4/38.9 MHz	(Intercarrier sound – 5.5 MHz)
"	PAL-I	33.5/39.5 MHz	(Intercarrier sound – 6 MHz)
"	NTSC	41.25/45.75 MHz	(Intercarrier sound – 4.5 MHz)

Figure 7.5

The tuner unit

The major function of this section of the receiver is to convert the RF signal on the aerial into the IF and introduce the minimum of additional noise and distortion; while at the same time maintaining the gain relatively flat across the channel bandwidth to retain the level balance between the luminance, chrominace and sound components of the signal. In the interests of low-noise mixing, it is necessary to ensure that the RF signal has an amplitude comparable to that of the local oscillator. Thus typically the conversion gain (RF input to IF output voltages) of the tuner will be in the order of about 30 dB. The aerial input to the tuner will be via an isolating unit typically consisting of a series capacitor of about 220–470 pF shunted to earth by a resistor of about 1–5 MΩ. This is used to prevent the aerial becoming live to mains voltages through inadvertent connection to the receiver earth line. The resistor acts as a discharge path for the capacitor that might well become charged through atmospheric conditions. It is important to use the correct plug and socket coupling between the feeder and tuner inputs and so maintain the impedance matching.

In practice, the tuner bandwidth is likely to exceed that of the wanted channel but it must be narrow enough to reject the image frequency. The gain and bandwidth up to transmission frequencies of about 1 GHz are now very easily achieved because transistors are readily available with gains as high as 12 dB at 12 GHz and produce less than 1 dB of excess noise. The typical UHF tuner has for a long time been constructed around components based on microwave waveguide principles, using trough line and lecher bar-tuned circuit elements. These are in fact made shorter than λ/4 and then tuned to resonance with the variable capacitance of varactor diodes using a stabilized bias supply that can swing from about 0.5 volts at the low frequency end up to about 30 volts for the highest channels. Currently, many receivers now employ digitally controlled frequency synthesis phase locked loops that include the feature of sweep or scanned tuning. In these cases, the receiver can be set into a mode that scans the frequency spectrum from the lowest channel upwards and halt when it detects a signal of suitable amplitude. If the viewer finds this signal to be the wanted channel and of acceptable quality, its channel data can then be stored in the system semiconductor memory.

Figure 7.6 shows the basic principles of frequency synthesis that leads to digital control over the receiver tuning system. The local oscillator signal is provided by the phase locked loop (PLL) which consists of a voltage-controlled oscillator, two dividers, a loop filter and a comparator. The device labelled ÷n is described as a prescaler which is used to extend the range of operation. The basic reference frequency is provided from a crystal-controlled oscillator for high stability and this frequency is divided down to provide an input to the phase

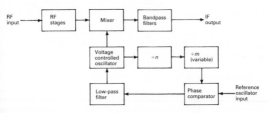

Figure 7.6   Basic phase locked loop tuning system.

comparator. If the reference oscillator frequency is 4 MHz and this is divided by 1024, the receiver oscillator will lock to multiples of 3.90625 kHz. When a channel is selected, control logic converts this into a frequency and the phase comparator causes the local oscillator to be adjusted until the output from the variable divider m, is equal in phase and frequency to 3.90625 kHz. If the prescaler had divided the local oscillator frequency by 16, the actual tuning steps will be 3.90625 x 16 = 62.5 kHz, a value well within the range of the system AFC system. Therefore by choosing suitable reference frequencies and division ratios, any tuning step size can be created. The local oscillator output can also be fed to a frequency counter that is offset by the IF value. The counter then provides a digital version of the tuned frequency. Because the division ratios are based on binary numbers, the concept can readily be adapted to digitally controlled tuning via a microprocessor. This then leads to the possibility of controlling the receiver functions via either the Inter-IC bus (I^2C bus) or the alternative three-wire bus system. The tuning of this type of receiver is very stable because all injection frequencies are derived from a crystal source. Problems that can arise include excessive noise due to jitter caused by the oscillator and spurious beat notes that are generated by the high speed synthesizer divider circuits.

Tuner AGC system

Because it is necessary for the tuner RF amplifiers to deliver a relatively large signal to the mixer input, it is important that the gain of the tuner should under normal signal levels remain high; certainly until signals of large enough amplitude to create overloading are encountered. The AGC action within the tuner should therefore be delayed until this condition arises. Because the IF section will be intended to produce a signal gain in the order of 45 to 50 dB, normal AGC action should be applied to this stage and used in such a way as to help maintain a good S/N ratio. The gain of a transistor amplifier is a maximum at some specific value of base or collector current and then falls as these currents are increased or decreased. This feature is employed in the automatic control of system gain. Forward AGC action occurs when the transistor is driven harder into conduction under large signal conditions (the alternative is described as reverse AGC). Forward-acting AGC is somewhat less sensitive than the reverse-acting system, but very much more linear in operation. To achieve this action, the transistor collector voltage is provided via an unbypassed resistor so that an increase in collector current causes the collector voltage to fall and reduce the stage gain. This feature is described as the Early effect. The increased collector current increases the damping on the tuned circuit amplifier load and tends to increase the bandwidth. This allows the large signal to be reproduced with good quality and frequency response, and at the same time swamp any extra noise due to the wider bandwidth. By comparison, reverse-acting AGC would narrow the bandwidth so that the large signal would be reproduced with a poorer frequency response. The presettable voltage delay to the tuner AGC circuit is normally obtained from the IF stage via a clamp circuit that becomes operative just before the IF signal is large enough to introduce overloading. An incorrect setting which results in overloading may produce excessive contrast, interference patterning or even a buzz on the sound channel.

Tuner fault conditions

Because of the frequency range involved and the component sensitivities to changes when the covers are removed, tuner unit faults tend to be dealt with on a replacement if found to be faulty basis. The repair of these devices is best left to specialists, because this invariably involves retuning and alignment, and the necessary test equipment is too expensive to be found in the general service workshop. Having made this point, there will be occasions when access to such service is not possible and then the service engineer will need to proceed with great care. Probably the weakest components within the tuner are the decoupling capacitors and the switching and tuning diodes. These, like the transistors, are prone to damage through atmospheric storms. Further difficulties for the service engineer are introduced by the use of ICs and SMDs (surface-mounted devices). DC tests on suspected faulty capacitors are not very reliable as high frequency ripple currents can easily generate a few hundreds of milliamps of current to create distress. Tests should be made at least up to the ripple frequency and preferably up to about 150 kHz.

For receivers using digital control, some apparent tuner faults can be traced to the digital circuits. For example, the receiver may tune manually or under scan control but fail to store the channel data. Alternatively there may be no sound or vision with the receiver stuck off channel. In these cases suspect a faulty power supply to the RAM or a faulty backup battery.

For the lack of tuning ability, check the varactor diodes' stabilized power supply, while tuner drifting should lead to a check for leaky capacitors on the tuning supply voltage line.

For a noisy picture with loss of gain, check the operation of the AGC system, and any preset control for poor contact.

Intermittent responses invariably point to dry joints in the control circuit.

PAL RECEIVER SIGNAL PROCESSING

The composite video signal is considered to consist of the summations of either the Red (R'), Green (G') and Blue (B') signals, or the luminance (Y') and the two colour difference signals (R'–Y') and (B'–Y') which when amplitude weighted (See Figure 7.3) become the V and U components of chrominance respectively. Both U and V components are amplitude modulated using DSBSC on the colour subcarrier frequency of 4.43361875 MHz during which process, the V signal is phase inverted on alternate video lines. This modulated component is then added to the luminance signal. Line sync pulses of 4.7 μs duration are added every 64 μs and a group of 5 broad field sync pulses each of 27.3 μs are added at the rate of 2 pulses per 64 μs. A burst of nominally 10 cycles of subcarrier frequency is gated into the black level of the signal between each line sync pulse and the beginning of the luminance. The active video signal has a duration of about 52 μs so that for about 12 μs after each line pulse, the signal is set to the black or blanking level. A further blanking period of nominally 20 line periods occurs at the end of each field. These resting periods, referred to as the horizontal and vertical blanking intervals (HBI and VBI), are provided to allow the system processing to settle between changes from video to synchronizing information and vice versa. This complex signal is often referred to as the composite video, blanking and sync (CVBS) and this is used to amplitude modulate the final carrier frequency.

It is thus the function of the receiver processing stages to recover the original R, G and B signals from this complex waveform. The block diagram of Figure 7.7 shows the stages of decoding that are used to recover these components. The waveforms shown in an idealized way in Figure 7.8 are intended to explain how this processing is achieved.

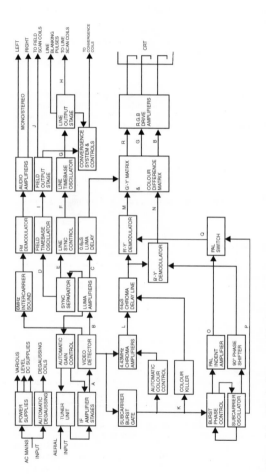

Figure 7.7 Block diagram of basic PAL system receiver.

154

(The input signal is assumed to be a modulated standard rainbow bar pattern.)

The waveform at A of Figure 7.8 represents a modulated signal as it will appear in almost all of the RF and IF stages, while that at B is the demodulated and filtered Y component. Since the wideband luminance signal passes through its amplifier stages rather faster than the corresponding chrominance components, it is necessary to balance this by introducing a short signal delay of about 600 ns. The waveform at C is thus exactly the same as that at B, but slightly delayed. At the sync separator stage the line and field pulses are processed through a differentiator and an integrator before being applied to synchronize their respective oscillator stages. The waveforms at D and E thus represent the recovered field and line sync pulses respectively, while that at F shows the amplified and clipped version of the line sync pulse.

Because of the large difference in operating frequencies, the impedance of the line and frame coils used to provide scanning of the CRT is almost entirely inductive and resistive respectively. The pulse-shaped voltage waveform shown at G thus generates the sawtooth current at H that is needed to provide line scan. Because the frame scan coils are practically resistive, a sawtooth voltage at I will generate a near sawtooth current through these coils. The actual current waveform shown at J is somewhat distorted due to the small inductive component of the coils impedance. The waveform at K shows the gated outburst of colour subcarrier that is used to lock the regenerated 4.433 MHz oscillator used for chrominance demodulation purposes.

The modulated subcarrier components shown at L, M and N represent the combined U + V component and the recovered R'–Y' and B'–Y' colour difference signals respectively. The swinging burst of the synchronizing colour subcarrier generates a signal at half line frequency that identifies a PAL transmission. This 7.8 kHz signal is used to switch the phase of the V demodulator subcarrier on a line-by-line basis. The two colour difference signals are input to a matrix circuit that regenerates the G'–Y' colour difference and then the Y' component is added to each to recreate the original R', G' and B' signals that are used to drive the CRT.

MULTISTANDARD DIGITAL DECODING

The digital processing of the video signal has a number of significant advantages, not only because it allows the introduction of such features as picture-in-picture and digital on-screen display of user control menus, but also because digital comb filters are much more effective in removing the cross-colour/cross-luminance artefacts than their analogue equivalents. It will be recalled that when high-frequency luma components leak into the chroma channel they create false-coloured patterning. In a similar way, leakage of chroma information into the luma channel produces busy or ragged vertical edges on an image. Ideally it would be an advantage to digitize the vision signal at the IF stage, but to process this wideband complex signal in this way is currently fraught with difficulties. The diagram of the IC shown in Figure 7.9(a) and (b) represents an example of a current typical compromise to this problem. It is quite easy to distinguish between the PAL, NTSC and SECAM signals by their chroma subcarriers and line and field frequencies. Both the PAL and NTSC signals use QAM for the modulation of different subcarriers and line and field frequencies. By comparison, PAL and SECAM differ only in the colour subcarrier frequencies which in the latter case use frequency modulation for two

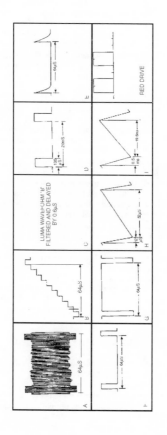

64µS

64µS

LUMA WAVEFORM 'b'
FILTERED AND DELAYED
BY 0.6µS

20mS

135
µS

RED DRIVE

64µS

64µS

5µS

19.9mS

0.6
µS

0.5
mS

64µS

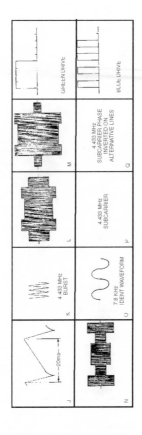

Figure 7.8 Waveforms associated with Figure 7.7.

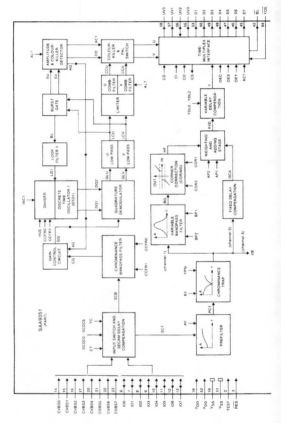

Figure 7.9 (a) Block diagram of digital multi standard TV decoder SAA9051 (courtesy of Philips Semiconductors).

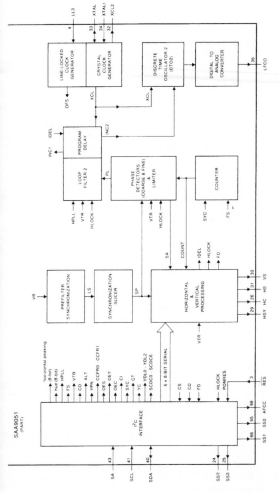

Figure 7.9 (b) Block diagram showing processing of PGLC and control signals SAA9051 (courtesy of Philips Semiconductors).

159

subcarriers. The inputs to this chip, a Philips Semiconductors SAA9051, are provided from a digitized version of the demodulated composite video signal (CVBS). If it detects a SECAM input signal it gates this out to a companion parallel digital processor. As development progresses, it is thought highly likely that the three chips, A/D converter, SAA9051 processor and SECAM decoder could ultimately be incorporated on to a single device.

The operation of the IC shown conveniently splits into two parts: that shown in Figure 7.9(a) is responsible for the signal processing while that shown in Figure 7.9(b) is related to the sync pulse and timing functions. The operation is based on the television standard of 13.5 MHz sampled data and the signals are input either via the CVBS or IO ports. The system operates in one of two modes for either Y/C or CVBS signal processing.

Y/C mode

The Y signal component plus sync pulses are input via the port $CVBS_0$ to $CVBS_7$ to the first stage, while the chroma is input through IO_0 to IO_7. The switching stage diverts the luma and chroma components through the appropriate channels (SCT and SCB).

CVBS mode

The 8-bit sampled composite signal is in this case directly input via the CVBS port to the first stage. Again the switching action separates the components into the luma (SCT) and chroma (SCB) channels. If a SECAM signal is detected, the control bit CT is set to enable tri-state logic buffers to divert the signal through the IO port to the parallel SECAM processor stage.

Luminance path

After filtering to remove the chroma signal, the luma is split into three channels. The control bit PF is used to set the prefilter either into the bypass mode where the gain response is flat, or to provide about 9 dB of HF gain. The control bits BY and YPN set the chroma trap to a centre frequency of either 3.58 or 4.43 MHz. At this point, the high-and-low frequency components of the luma signal are processed separately. The high frequencies of the channel 1 signal are passed through a programmable variable bandpass filter whose centre frequency is set to one of four values, ranging from 2.6 to 4.1 MHz through the two control bits BP_1 and BP_2. The signal BG is passed through the corner correction or *coring* stage which removes any low amplitude noise. The degree of noise reduction being controlled via the two bits COR_1 and COR_2. The low-luma frequencies in the channel 2 signal pass through a delay compensation circuit before being input to the combining stage. The HF is now weighted by one of four levels which are determined by the control bits AP_1 and AP_2 before being added to the delayed component DCA. The signal AVD is then passed through a variable delay circuit to match the transit time delay of the chroma signal through its processing stages before being input to the output multiplexer interface. Channel 3, VB carries the synchronizing signals to the timebase section of the chip.

Chrominance path

The chroma signal component SCB is passed to the chroma bandpass filter whose characteristics are set by the subcarrier select control bits

CCFR$_1$ and CCFR$_2$ to process either PAL signals with subcarriers of 4.33 MHz or 3.575 MHz, or NTSC or PAL signals with 3.58 MHz subcarriers. The selected chroma signal CG is passed through the automatic colour gain control stage which is under the control of signal AG developed from the amplitude of the burst signal in the colour killer detector circuit. The signal GQ and the two quadrature signals from the oscillator DTO$_1$ are input to the demodulator stage to recover the U and V chroma components. These are low-pass filtered and limited before being input to the colour killer and PAL switch stage. The 4.43 MHz colour burst signal is gated out to the colour killer detector and after being filtered it is used to lock the regenerated colour subcarrier. This oscillator frequency is preset by the 2-bit inputs from the CCFR$_0$ and CCFR$_1$ signals and then synchronized by the input INC$_1$ derived from the TV/VCR detect circuit. The user HUE control can vary the colour subcarrier through ±180°. In the PAL mode, the colour killer and PAL switch stage restores the correct phase of the V signal which is then input together with U component to the output interface.

The multiplexer stage then outputs the Y signal through the port D$_1$ to D$_7$ while the chroma output is processed as alternate 4-bit groups of U and V signals via the port UV$_0$ to UV$_3$.

The remaining control bits have the following functions:

AC$_1$	colour killer status,
BL	blanking line, active low,
CD	PAL/NTSC detect,
CI, CO	colour on/off status,
CS	SECAM colour detect,
FOE	enable/disable – active low for picture-in-picture feature,
OEC	chroma output enable,
OES	sync output enable,
OEY	luma output enable.

The third channel signal from the luma chain (VB) is filtered and shaped to provide the sync pulse train SP which is input to the line and field timing and processing stage. XTAL and XTAL$_1$ inputs provide an accurate crystal clock input of nominally 24.576 MHz which are used together with oscillator DTO$_2$ and the sampling clock rate input LL$_3$ running at 27 MHz. The frequency of 24.576 MHz is chosen because it is a common multiple of the PAL and NTSC line frequencies (15.625 kHz x 434 = 6.75 MHz and 15.734254 x 429 = 6.75 MHz). The sync and processing systems are thus locked to line frequency in both modes. DTO$_2$ also provides an analogue output LFCO to drive the parallel SECAM decoder chip. The crystal signal is time delayed and used via the loop filter to lock the PLL system. This is under the control of a divider circuit driven by the counter. The timing of the vertical (VS) and horizontal (HS) sync pulse outputs are thus very strictly controlled. Like oscillator DTO$_1$ in the signal processing stage, oscillator DTO$_2$ is also controlled via a time delay INC$_2$, both of which are generated from the master delay signal IDEL. The I^{2}C bus interface circuit is controlled via the SCL or clock line, with the SDA line providing the serial data. The signal SA provides a selective address.

The remaining control signal functions:

AFCC	additional output control,
FD	field frequency detect,
FS	60/525 or 50/625 system select,
HC and HSY	sync mode status,
HLOCK	horizontal line lock status,

HPLL	line frequency clock,
OFS	field sync output,
PONRES	power on reset,
SS_0 to SS_3	source select,
VTR	TV/video select.

Possible problem areas

Because this chip relies very largely on other highly integrated circuits, there are few external components to create fault conditions. Most faults are therefore likely to be associated with printed circuit track problems such as open or short circuits causing lines to be permanently jammed at logic 1 or 0. While most of the signal status bits are hidden internally, a significant number of useful faultfinding indicators are available at the various pin connections. Apart from the digital input and output versions of the television signal, the blanking and sync pulses (BL, HS, and VS) are readily available for display on an oscilloscope. Likewise, the clock signals associated with the LL_3 and XTAL pins can easily be compared for incorrect timing. If the FOE line happens to be jammed in the high state, output of the Y, U and V signal components will be absent.

FIELD/FRAME SCANNING STAGES

This stage runs at a fixed frequency of 50 or 60 Hz to produce a sawtooth field scan current. Because of this low frequency, the field scan coil impedance is very largely resistive but with a very small series inductance. Thus when a sawtooth voltage is applied across the circuit it generates a sawtooth current, but with a very small distortion component which can readily be eliminated using negative feedback. The power requirements and frequency response of this circuit are closely akin to those of an audio amplifier system; however, the negative feedback loop is used to generate the distortion to the sawtooth waveform rather than to correct it as in the audio case. Furthermore, the scan coil impedance is usually included in the feedback loop in order to compensate for any loss of height that might occur as the temperature of the coils rises. The typical field scan system is included on a single chip requiring only DC supply and field sync pulses as inputs. The power output stage commonly consists of a Class AB or Class B push-pull circuit. The 20 ms periodic field sawtooth waveform is divided approximately into 19 ms forward scan and 1 ms flyback. The timebase oscillator is commonly based on charging a capacitor via a low-value resistance (often a forward-biased diode) to the supply rail value of about 25 volts. During flyback, the diode becomes reverse biased and rapidly discharges the capacitor. Because the vertical height of the screen is relatively small, the pincushion distortion effect is much less important than in the case of the line scan system so that S and N-S correction is often ignored. Because loss of field scan would create a bright horizontal line across the tube face which could burn the phosphors, field scan ICs usually carry a detector circuit that shuts down the tube supply voltages under fault conditions. This obviously creates a problem when faultfinding in the scanning circuits.

Figure 7.10(a and b) show a block diagram of the construction of such an IC, the Philips Semiconductors TDA3654, and the way in which it is connected into circuit. This chip provides an output that is capable of driving either 90° or 110° tubes with a peak-to-peak current of 3 amps. The internal voltage stabilizer circuit provides a stable output current even when the supply voltage varies. The guard circuit

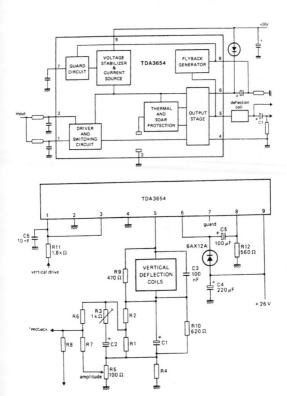

Figure 7.10 (a) Block diagram of field processor IC. (b) Application diagram of processor (courtesy of Philips Semiconductors).

provides protection against field scan failure and the thermal protection circuit ensures that the chip operates well within the safe operating area of the device characteristics. The output stage consists of a pair of Darlington amplifiers operating in Class B. A field sync pulse input of 3 volts amplitude at a total current of only 3 mA will fully drive the output stage. During the scan period, capacitor C5 (See Figure 7.10(b)) charges through R12 to the supply voltage level and flyback is initiated when the voltage at pin 5 just exceeds this value. The diode connected to pin 5 ensures that C5 is rapidly discharged. The resistors R3 and R5 in the feedback loop are used as linearity and height controls, with C2 providing a degree of S corrections for flat-faced CRTs. Loss of height and linearity problems should direct attention towards the electrolytic capacitors which can dry out and become leaky. A further problem can arise with the complementary push-pull output stage, particularly with the discrete component circuits. Here the audio equivalent of crossover distortion can give rise to cramping of the middle band of horizontal lines again; leaky capacitors are often the culprits.

Line scan stage

This stage is almost invariably designed around a flywheel sync system which incorporates a PLL circuit. The system low-pass filter loop has a relatively long time constant in order to average out the effects of the broadcast line sync pulses which may be affected by impulsive noise and interference. For use with a VCR or video disc player, this time constant may be shortened because these line sync pulses are more likely to be affected by jitter from the player switching system than by noise. Hence this filter time constant may well be one of the many compromises made in receiver design. Figure 7.11(a) shows the basic organization of this stage. Because the line scan coil impedances are almost entirely inductive, a square-wave input voltage will generate the necessary sawtooth current through the coils due to the integration effect of the inductance at the higher line scan frequency. Thus the switching input waveform to the line output stage (see Figure 7.11(b)) produces a sawtooth current through the scan coils L1. The output stage behaves much as a resonant circuit with energy being continually transferred between the inductive and capacitive components of the circuit, but oscillating at different frequencies on each half-cycle. The output transistor TR1 may be an NPN bipolar type or MOS field effect device, but must be able to withstand a voltage as high as 1500 and support a current of up to 6 amps. In order to

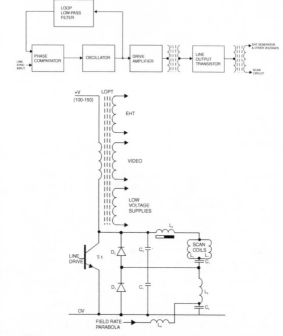

Figure 7.11 (a) Block diagram of line timebase circuit; (b) Basic line output stage.

maintain a low level of heat dissipation, the device must be switched very rapidly from one state to the other. Typically, the maximum operating temperature will be in the order of 150^0C. It is worth pointing out at this stage that the line output transistor for a computer monitor using apparently similar conditions but with a 64 kHz line frequency will need to dissipate perhaps twice as much heat as the TV receiver timebase running at about 16 kHz. Because of the power output needs, the driver transistor must itself produce about 3 to 5 watts of power, to maintain a drive current that can be as high as 3 amps. To some extent, this is caused by the fact that Tr1 does not turn off when the base voltage falls to zero due to a storage effect. It requires a reverse current to discharge the base capacitance to achieve the rapid switching action in less than about 2 μs.

Starting with zero current flowing in the LOPT primary winding (the load), the scanning beam positioned at the tube face centre and Tr1 just turned on. This action short circuits the components associated with D1, D2, C1 and C2. The switching action suddenly connects the supply voltage across the primary of the LOPT winding so that the current through it now builds up in a linear ramp fashion. After a period when the current is reaching a maximum (at the end of the scan), Tr1 is rapidly turned off. At this point, the current flow reverses and the energy stored in the inductor now transfers to the capacitive components, but at a much higher rate. Up to this point, the changing current has created a scanning action, first driving the CRT beam to the tube face edge relatively slowly and then by a rapid flyback action so that the beam is once more at the face centre, but with energy stored in the capacitance. This energy is again returned to the inductance but at the continuing high retrace rate, the inductive current falls relatively slowly to create the initial part of the scan stroke. In a typical 64 μs scan period, the half-cycle scan and retrace periods are approximately 52 and 12 μs respectively. The diodes D1 and D2 together with capacitiors C1 and C2 form part of the energy efficiency circuit, adding scanning energy when that from the main circuit is zero. These diodes also help to switch the circuit between the two resonant frequencies. The small positive resistance component associated with the scan coils can create distortion. This is countered by adding a negative resistance in the form of a small saturable inductor in series as shown, but with a variable permanent magnet to allow for circuit linearity adjustments. In order to improve the regulation of the EHT voltage, the LOPT is often tuned to the third or fifth harmonic of the line frequency.

Because the tube face and beam deflection angles differ, the linear scan will be distorted. This is corrected by the addition of the series S correction capacitor C1. Again for the same reason, the scan length linearity from top to bottom of the screen would also be distorted and this is corrected by the E–W correction circuit. This involves modulating the line scan width with a parabolic waveform obtained from the field timebase, via L4, L3 and C3 in Figure 7.11(b).

Line output stage faults

The circuit provides voltages for most of the other stages in the receiver and is commonly linked to the switched-mode power supply section. There is therefore a high degree of interdependence with this stage. Problems of mains power supply stability can lead to variation of scan amplitude (width) and changes of EHT demand as the brightness varies. Because of the presence of very high voltages and currents, the circuit is prone to dry joints which in turn can develop into component failures. The trick of disconnecting this stage from the mains power and replacing its load effect with an incandescent lamp bulb can be used to advantage here. If after replacing a defective line output transistor, the receiver works but with some image distortion, then suspect a further component that could modify the collector current. A change in parameters over time and after stress can easily create a failure in an associated device such as a leaky capacitor, diode or even an obscure dry joint. After changing a line output transistor (or any heatsink-mounted device), it is important to ensure that the mica and insulator washers are in good condition and that new compound is applied to both sides of the mica. Finally, ensure that the device is secured with its screws to the heatsink, but not over-tightened.

Colour drive circuits

The major requirements for the large signal amplifiers in this stage are a high gain bandwidth product, a high slew rate and a gain in the order of 35 dB. These parameters can readily be met using an ASIC device such as that shown in Figure 7.12 (courtesy of Philips Semiconductors). This device has a gain bandwidth product of 750 MHz, a large signal bandwidth of 7.5 MHz and a slew rate of 1600 V/μs. To maintain a low self-capacitance and protection from CRT flash-over, the chip is intended for mounting on the tube base circuit board. The major advantage of such a chip is that the parameters of the three separate stages red, green and blue can more easily be matched in production. The device is constructed as a simple 9-pin in-line chip running from a common supply voltage line of about 200 volts and the cathode drive current is limited to 5 amps peak. CRT flash-over protection is provided internally by a diode that clamps the outputs to $V_{DD} + V_{Diode}$ and externally by series 1k5 carbon resistors in each drive line and these are in turn shunted by 2 kV spark gaps.

The device parameters are carefully controlled through the use of current-mirror circuits in both the V_{DD} and GND supply lines where the output current follows closely or mirrors the input current. These are basically amplifiers with very high output and very low input impedances and help to maximize the gain bandwidth product.

Figure 7.13 shows a typical application of the TDA6103Q chip. The method of setting up these stages varies with manufacturer but the following is a good approximation of the general approach using a

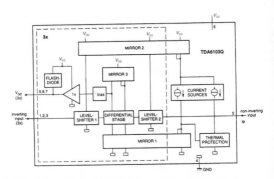

Figure 7.12 Functional diagram of RGB drive IC (courtesy of Philips Semiconductors).

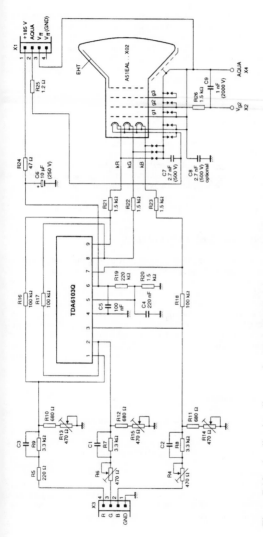

Figure 7.13 Application diagram for TDA 6103Q (courtesy of Philips Semiconductors).

167

standard colour bar pattern. The resistors R13, R14 and R15 are used to set the black level and R4 and R6 are used to set the gains of the green and blue channels to balance that of the fixed gain red channel. Turn off two beams at the CRT base (say green and blue), set the black level control for the remaining gun to near mid-position and adjust the A1 voltage level (obtained from the potential divider network of the focus control) to obtain the just cut-off condition. Repeat this in turn for each of the other guns, but only adjusting the black level controls to achieve cut-off. Check that the A1 voltage lies within the specified range and readjust if necessary. With all three guns in operation, adjust the brightness and contrast controls to provide bars with a graduated scale between black and white. Advance the colour control to obtain the correct colour bar balance, but recheck this on a normal picture using flesh tones as a standard.

Fault conditions

Problems in this stage of the receiver usually involve the loss or reduction in gain of one channel and if the standard colour bar pattern is used, it is easy to detect where the cause of the fault lies. One peculiar fault that has been reported on occasionally results from replacing a CRT with a regunned device. When set up and installed, most of the colours had been found to be incorrect. The cause was traced to a gun assembly with different red, green and blue pin connections.

COLOUR TUBE PERFORMANCE

Historically, the ability to achieve a situation whereby all the colour beams land only on their respective pixel phosphors has called for some dextrous manipulations by the service engineers. All colour CRTs use some form of mask or grille structure to ensure that as far as possible this condition is always met. Because of this, the mask actually gathers some electrons from the beams which results in a mask current that creates heat and hence mechanical distortion, leading to spurious colourations. Beam current limiting helps to minimize this effect by monitoring the average current drawn by the EHT system or the tube cathodes. This feature is used as a control to either turn down the tube drive by negative feedback applied to the luminance stage or to shut down the switched-mode power supply under worst case conditions.

Purity adjustments

This operation is needed whenever the three beams spill over on to neighbouring pixels to create spurious colourations. Switch off blue and green guns, slide back the scan coil assembly and adjust the orientation of the purity magnets to produce an ill-defined red blob at the tube centre. Slide the scanning yoke forwards until a complete red raster is obtained. Check with only the blue and green guns on separately to ensure that these are also pure. Any discrepancy can be eliminated by a compromise adjustment of the purity magnets and the position of the scanning yoke. A good indication of the effect of the earth's magnetic field can be demonstrated by turning the receiver upside down.

Electron beam convergence

This operation involves individually deflecting the electron beams so that a white raster can be created over the whole screen. Ideally this operation can be best carried out using a cross-hatch or dot pattern or a test card. Static convergence controlled either by permanent magnets or by DC energized coils affects the central image only. Dynamic adjustment of the wider screen area is affected by bleeding controlled levels of line and field rate parabolic current waveforms into the opposite scan coil pair.

Pincushion distortion

This problem shows up as bent verticals or horizontals in an image and is particularly troublesome with large screen tubes. The extent of the problem is readily seen when displaying a cross-hatch pattern. Again this can be corrected by applying a controlled amplitude-modulated waveform from line to field and field to line scanning systems.

Modern CRTs

Many are now supplied with matched scan coil and convergence assemblies so that no form of adjustment is needed, which is particularly advantageous for wide screen displays. Even pincushion distortion is avoided by the supply of pin-free tubes.

LIQUID CRYSTAL AND LIGHT EMITTING DEVICES (LCDS AND LEDS) AS APPLIED TO SOLID STATE IMAGE DISPLAYS

Portable TV receivers and computer monitors are now readily available with solid state electronic display panels with up to 42 cm diagonal size and based on either liquid crystal or light emitting diode technologies. Although much research is currently in progress to find a suitable alternative for the CRT, these two devices are presently the only real alternative. They are much thinner, lighter and more robust than the equivalent sized CRT. While these devices operate on different principles, the pixel locations which are sited at the intersections between vertical and horizontal sense lines are addressed in a very similar manner. The principles and behaviour of LCDs and LEDs are generally well understood and so these will not be described here. While the LCD controls the intensity of energy output from a back light source through a matrix of colour filters, the LED generates red, green or blue light directly from an array of diodes. In both cases, the array of filters and diodes are arranged in a triad matrix formation, in the same way as for the colour CRT. Each pixel cell may be switched on or off via thin film transistors (TFT) that are also fabricated directly on to a glass faceplate. To give an example of the size of each cell, a typical 14 cm diagonal display device will be fabricated with 640 × 480 × 3 individual pixel elements, with each being capable of creating 256 variations of light level. The selection of a particular pair of row and column address lines then selects that pixel for illumination. Raster scanning and pixel control is achieved by using a controller chip built into the display panel. This converts the line and field sawtooth waveforms into the appropriate row and column selection pulses. At the same time, a pulse whose amplitude represents the level of the video signal is multiplexed on to one of the lines to provide the correct degree of contrast for the selected pixel. In normal operation, both of these devices are very reliable and any fault is likely to involve the complete replacement of the panel and its controller chip. The loss of one cell in an LCD or LED device generally shows up as a permanent bright or black spot respectively.

DIGITAL CONTROL OF THE TV RECEIVER

The availability of low-cost microprocessor chips has significantly changed the methods of control of both TV receivers and video recorders. In fact, this market has spawned a new small area network system referred to as the Inter-IC bus (I^{2}C bus). Because only two lines are dedicated to control, the cost in terms of pinouts on each IC is small. The two lines provide for bi-directional serial data transfers at rates up to 100 kbit/s plus a clock signal. The bus operates on a master/slave basis but allows for several devices to act in turn as master, each providing its own clock signal. An arbitration procedure prevents bus conflict between devices, thus allowing only one master to gain control at any one time. Each IC on the bus has a unique 7-bit address which is contained in the first byte of each data transfer. Each device compares this byte with its own address and only if a match is found will it respond to the master that transmitted the address. The number of data bytes that follow is unrestricted and the slave acknowledges the received data on a byte-by-byte basis.

An alternative three-wire bus is used by some manufacturers and the lines cater for data, clock and device enable, with the latter controlling the communicating pair of devices. The data block is normally a single word carrying information and port address which is transferred only during an enable line high period.

With such a communications system it is very easy to add diagnostic information for service purposes. The system is controlled either via the remote control infra-red (IR) handset or a dedicated service controller. Such a device provides a valuable field service tool, allowing the faulty circuit area to be quickly located and the circuit board changed. However, this can produce extra wear and tear on the board edge connectors, leading to possible future problems.

The service mode can be entered in one of two ways, either via the normal handset or the service controller, or by earthing one specific pin or service point on the receiver chassis. The service type handset has a further useful advantage, it can be pre-programmed with all the normal local area channel data, plus the settings for normal use, thus making for ease of installation. In operation, these systems either use the CRT or VCR display panel to show the results of testing. The response to a particular test yields a set of codes, usually in hexadecimal, which with the aid of the service manual points the engineer to the faulty stage and even suggests the possible faulty component. In some cases, the system has an error code memory of up to 10 registers which can be helpful in diagnosing intermittent faults. The fault codes thus form part of a diagnostic algorithm. Such a system could not possibly be without its own faults and of those reported, the most consistent relate to typical digital problems of control lines stuck at logic 0 or 1. For example, no on-screen graphics should lead first of all to check around the microprocessor. With tuning stuck on one channel, again suspect the microprocessor stage, but at least in this case the engineer has an on-screen display. Incorrect tuning selection can often be resolved by clearing the memory and reprogramming the system.

Infra-red handsets

While these devices are generally very reliable, they are also prone to damage through being dropped. A simple test to check the operation is to aim the device at the aerial of a portable long or medium wave band receiver when interference will be heard from a normally working handset. If the receiver fails to create a response, the drive to the LEDs can usually be checked with an oscilloscope to determine the state of these components. If there is no output, then checks should be made of the battery connections and condition, any large-value capacitor used to store the energy to drive the handset and the possibility of dry joints on the PCB. Failure of a single popular function is usually caused by contact wear on that particular button and in this case it is sometimes possible to provide a temporary repair by shifting the function to another button, or even turning the switch membrane so that a relatively little used function is placed in this position. Generally, the replacement cost of these devices does not allow for any significant repair. If the handset is proved to be serviceable then attention needs to be directed towards the receiver control panel where the IR detector stage should first be checked. Faultfinding in the control receiver section then follows normal digital techniques.

TELETEXT SYSTEMS

The World Standard Teletext (WST) system was originally devised for 625 line PAL systems but has also been modified to run on 525 line NTSC systems. For the 625 line PAL systems, there are 25 inactive lines in each field period, while for the NTSC system, the corresponding number is nominally 20. These unused VBI lines are therefore chosen to carry this digital information service as a time-multiplexed signal component. Each *screen, frame* or *page* is designed to contain 24 rows, each carrying 40 characters or graphics symbols. The pages are organized into *magazines* nominally containing 100 pages, with up to 8 magazines per transmission channel.

The data for each row is synchronously transmitted during the period of one VBI line. Each character or graphics symbol cell in the display is defined by six dots or pixels per line and 10 lines deep in each of the two interlaced fields. Each character and row is separated by one dot so that in total each cell represents an area of 5 x 18 rectangular pixels. For graphics representation, the cell borders are omitted and the characters are then said to be *contiguous*. The page information is displayed during the middle 40 μs period of each TV line. The colour palette of the display for the original Level 1 system is limited to three saturated primary (red, green and blue) and secondary (cyan, magenta and yellow) colours, plus black and white. However, developments that will be explained later will cater for very much more variability.

Each data line carries the binary digits as a two-level, non-return-to-zero (NRZ) code. The pulses are filtered and shaped so that most of the energy is contained within a bandwidth of about 5 MHz. The timing and signal levels relative to the peak white and black levels are shown in Figure 7.14.

The data bit rate is chosen to be an even multiple of the line rate and for the 625 line system the data clock runs at 6.375 Mbit/s (444 × 15.625 kHz). The corresponding bit rate for the 525 line system typically lies in the range 4.0 to 5.5 Mbit/s.

The teletext data signal, which consists of 45 bytes per line, is multiplexed into the period of about 52 μs, between the colour burst and the following sync pulse in the manner shown in Figure 7.14. Odd parity is used for error control, to ensure that there is always at least one data transition in each byte. This improves the synchronism of the decoder bit rate clock.

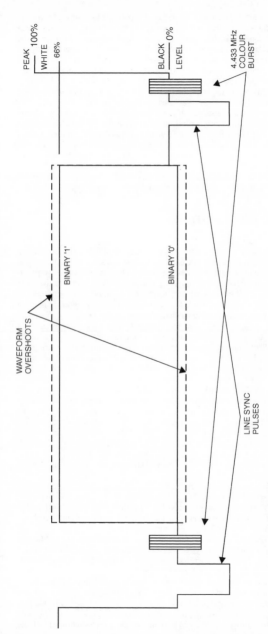

Figure 7.14 Teletext data levels relative to VBI.

The structure of the two types of data line is shown in Figure 7.15. Each line carries 360 bits in 45 bytes and each line begins with three identical bytes; two bytes of alternate 1s and 0s, termed the *clock run-in*, followed by a *framing byte* of 11100100. These are provided to synchronize the bit rate clock and to identify the beginning of each block of data. The fourth and fifth bytes are described as the magazine and row address group and are used to uniquely indentify a row in any particular magazine. Three bits are used to identify the particular magazine and five bits to identify a maximum of 31 rows. The following two bytes occur only in the *header row* and identify the particular page number within a magazine. All of these bytes are Hamming code protected so that all single bit errors are correctable and double errors detectable.

The header row also carries four bytes that provide a four digit time code (not necessarily related to real time) and by invoking this sequence, up to 3200 different versions of the previously identified page can be selected. Therefore a particular page may be identified by its magazine, page and time code. The following two bytes contain 11 useable bits for the system control. The header row ends with eight bytes which are used to provide a real-time clock display on each page. Since each displayed line occupies 40 µs and carries 40 characters, each using six horizontal pixels or dots, the display or dot clock must run at 6 MHz.

The character code set that is commonly used is shown in Figure 7.16. This is referred to as the ISO-7 code and is a variation of the standard ASCII code set. The bits for each symbol are transmitted in the order b1 to b8. This code format, which provides for 96 alphanumeric and 32 control characters, can be extended by dropping the parity check on character bytes. The use of eight bits then provides for an extended character set of 192 alphanumeric and 64 control codes, which allows characters from different languages to be displayed.

Fastext or full level one features (FLOF).

Each teletext page requires $24 \times 45 = 1080$ bytes of memory, so that page memory is often allocated into blocks of 2 kbytes per page. Typical current decoders can store the data for anything from four pages up to 200 using relatively low-cost memory chips. As explained above, the row address group includes five bits that can be employed to select 1 of 32 (0 to 31) different rows, while the page structure only requires 24. These additional rows can therefore be used for service extensions.

Packet 24 (prompt packet) carries keywords that briefly explain the nature of the information on the first four linked pages. This is displayed as row 25 but with each keyword colour coded. The indicated page is then selected by the user operating a single key of the appropriate colour.

Packet 26 can be used with a microprocessor-controlled system to provide additional characters to extend the basic language capability and provide additional non-teletext functions.

Packet 27 (link packet) provides a link between pages with a similar context. These pages are then automatically stored in memory and can be accessed by the user with a single key stroke when needed.

Packet 28 may provide a page key for the descrambling of encrypted data contained in packets 1 through to 25.

Packet 29 has a number of designated features ranging from an extension to the functions of packet 28, to character set designation.

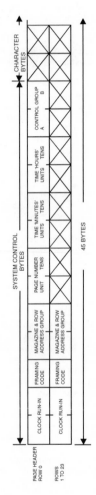

Figure 7.15 Teletext page code structure.

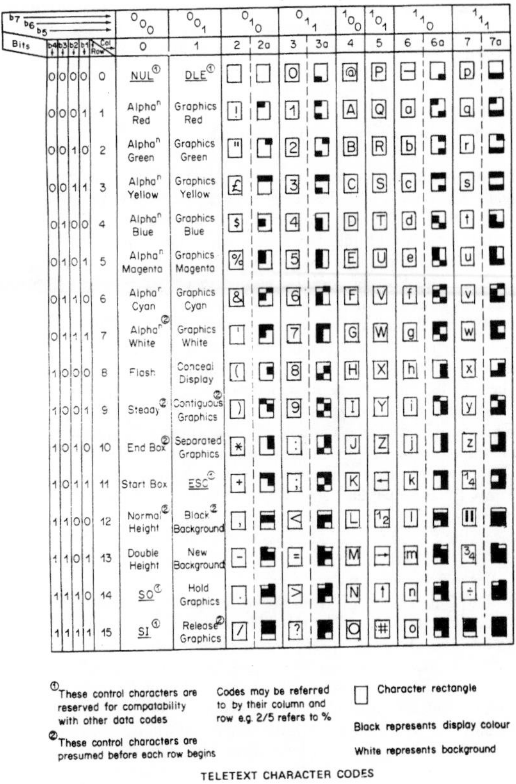

Figure 7.16 Teletext character codes.

A packet with the address 8/30 (magazine 8, row 30) carries designation codes that allow subaddressing which reduces the access time after the first page acquisition.

Packet 31 may be used to provide data services entirely independent of teletext.

Top of pages (TOP)

This is a technique that functions in a way similar to Fastext. The page linking information and page titles are transmitted on dedicated pages rather than as extension packets. This concept divides each magazine into blocks, groups and direct access pages, such as football, information and the results.

Various further system levels have been defined in proposed standards documents as follows:

Level 1.5. As FLOF but with an extended character set.

Level 2.5. Each page contains up to 32 colours selectable from a palette of 4092. Although designed for 16:9 displays, this mode is compatible with 4:3 receivers that use Levels 1 or 1.5. This is possible because the first 40 characters in a line are used for text, while the others may be used for graphics that are displayed as side panels.

Level 3.5. This extends the number and complexity of the redefinable characters and also introduces different font styles.

The Texas Instruments UNITEXT decoder

For more than 20 years, Texas Instruments have been producing teletext decoders and this latest version is a good example of the contraction that can be achieved by using ASIC chips. Basically the system consists of just two chips, the decoder and an analogue interface, but in practice this may be extended by the addition of extra page memory. Furthermore, a separate reset chip may be used to ensure that any reset command for other microprocessor-based devices on the I^{2}C bus do not interact with the teletext controller.

The operation of this single page decoder that caters for almost all European languages can be explained using Figures 7.17 and 7.18. The micro-coded processor handles all the teletext processing plus any other on-screen display (OSD) functions that may be needed. System control is via the I^{2}C bus where the decoder functions as a transceiver with the slave address of 0010001 or through a simple serial link CEN. The I^{2}C bus with its standard protocol can transfer data at up to 100 kbit/s. The CEN line which acts as a unidirectional multiplexed bus is provided for those receivers that do not employ the I^{2}C bus. In these cases, the SDA line is used to *assert* the data on the CEN line while the SCL line provides the clock signal.

The interface chip recovers the data from the VBI window included in lines 2 to 22 and 314 to 334 and regenerates the bit clock but at twice the rate of 13.875 MHz. The serial data is clocked into the decoder via the Prefix processor where it is converted into parallel format before being error checked and then passed through the micro-coded processor into the display memory under the control of the memory access controller. This stage arbitrates the access to the memory to avoid conflicts with other OSD needs. The character generator ROM provides the pixel output data through the video diplay processor which also generates an I signal to provide an *insert function* for mixed-mode display or boxed subtitles. All of the system timing functions needed for dot clock, text display and field sync are generated and controlled by the PLL.

The EUROTEXT Decoder

This later derivative of teletext decoder by Texas Instruments represents an intelligent decoder with inbuilt micro-coded processor to capture and store up to eight pages of data within its on-board RAM. It is FLOF compatible and provides for TOPs, favourite page accesses and programme delivery control for VCR operation through Packet 26 decoding. The decoder which is shown in Figure 7.19 alongside the UNITEXT decoder with which it is pin compatible, supports all east and west European languages including Cyrillic, Hebrew and Greek. This latest decoder is equiped for wide screen display (16:9 aspect ratio) through the use of switching linked to Line 23 data.

Faults in teletext receivers

The teletext service provides a good example of the advantages of a digital system. While the television image quality will degrade gracefully as the signal-to-noise ratio falls, the teletext signal remains solid

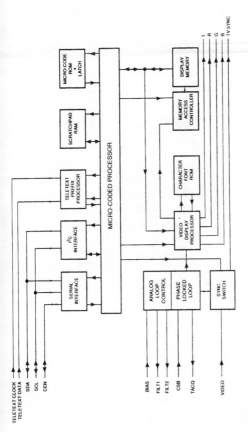

Figure 7.17 System level block diagram for the UNITEXT Decoder (courtesy of Texas Instruments).

177

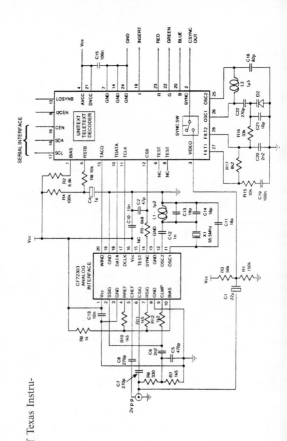

Figure 7.18 Applications circuit – (courtesy of Texas Instruments).

178

Figure 7.19 Teletext boards: (a) daughterboard; (b) UNI-TEXT board; (c) EUROTEXT board (courtesy of Texas Instruments).

until the service crashes at a relatively poor S/N ratio. Therefore, in general, this part of the service is the most reliable. The most common faults appear to be associated with handsets and board edge connectors. Unlocked or erratic text display may be caused by faulty phasing of the 6 MHz dot clock circuit. Loss of teletext service can result from failure of the 6.9375 MHz bit rate clock. A Teletext ROM or character generator fault can produce a constant error in the same character throughout the display, while a RAM fault can be seen as a consistent display error in the same page position.

A *clock-cracker* page can be a valuable asset when faultfinding in both analogue and digital parts of the TV receiver. At the time of writing, such a page which consists of a header row with real-time clock plus a full page matrix of alternate division signs and blank white character rectangles is present in ITV page 699. Generally, if the TV signal level falls, the real-time clock gives the first indication of problems. However, any errors in the major part of the page can be readily counted at each first page acquisition. This information can be used to gauge the quality of the vision signal as well as that of the teletext service. It might be useful to state here that since the clock cracker page is in monochrome, it can in an emergency be pressed into service as an aid in the setting up of the display convergence circuits.

CHAPTER 8

SATELLITE TELEVISION SYSTEMS

In simple terms, the satellite direct to home (DTH) television system utilizes the satellite as a relay station mounted in space, and to simplify the reception part of the system the satellite must appear to remain in a fixed position. This is achieved by locating the satellite at an altitude of about 36 000 km above the earth's equator and with a west to east velocity that is slightly greater than 3 km/s. Each satellite carries a number of components described as transponders which simply convert the received uplink signal frequency from a ground station into a new frequency for re-radiation back to the earth-bound receivers. Because such a satellite appears to be stationary to an earth-bound observer, this is described as a *geostationary orbit* (the Clarke orbit). Located in such a position, each satellite can *see* roughly one third of the earth's surface within an antenna beamwidth of about 17.5°. Although the signal path length will nominally be around 40 000 km in each direction, the propagation conditions are practically constant and predictable. In practice, a typical *earth footprint* covered by a satellite has an antenna beamwidth of about 2° north–south and 1° east–west, thus concentrating the signal energy into a relatively small region. For example, a transponder with a power output of 150 watts and an antenna with a gain of 43 dB, provides an effective radiated power (ERP) of about 3 MW. Such a system then becomes economically viable when compared with the costs of establishing a completely new terrestrial service. The only blind spots in the reception area are caused by high-buildings or high-sided vehicles, trees and forest areas and a latitude location above about 80° north or south. Just for once, the installation height of the receiving antenna is of little consequence; in fact, the lower the antenna then generally the shorter will be the feeder cable with a lower degree of attenuation.

The up-link transmission bandwidths typically lie in the ranges of 14–15 GHz or 17–18 GHz with a transmitter output power in the order of 150 W. This is fed to a parabolic reflector antenna of about 8–10 metres in diameter which achieves a gain of about 60 dB. The down-link frequencies lie in the ranges of 10.7–11.7 GHz and 11.7–12.5 GHz. (These frequency ranges form part of the Ku band.) The former is referred to as the FSS (fixed satellite services) band which was initially allocated to cable system head-end feeds and telecommunications services, while the latter is known as the BSS (broadcast system services) system that was initially allocated to direct broadcasting by satellite (DBS) the forerunner for DTH. Outside of Europe, the so-called C Band frequency range is often used and this lies between 3.75 and 6 GHz.

There are a number of economic and technical advantages in using the high section of the band for uplinking. The gain of an antenna is proportional to the operating frequency and inversely proportional to its beamwidth. If the higher frequencies were used for the downlink, any given antenna would have a narrower beamwidth, thus increasing the problems of antenna alignment as the satellite drifts around within its nominal mean position. This problem could only be resolved by using a larger receiving antenna or by adding a steering control to the receiver system. By using the high band for the uplink, the extra gain can usefully be employed to make up for the path length attenuation, losses which increase with operating frequency. With this arrangement, the overall system S/N ratio can be better managed. The large dish ground station will in any case require a servo-controlled tracking system, so there is no added cost in this case.

The total bandwidth allocated to each satellite is shared between its transponders so that the actual channel bandwidth depends to some extent on the number of transponders carried by the satellite. For example, the ASTRA system operated from Luxembourg by the Société Européene des Satellites (SES) has the capability of controlling up to eight co-located satellites at 19.2°E. The four satellites 1A to 1D each have television capable transponders with channel bandwidths of 29.5 MHz which are then organized in an overlapping manner using alternate vertical and horizontal polarizations, with carriers actually spaced by 14.75 MHz. ASTRA 1F and 1G, which have been designed to handle digital transmissions, have transponder bandwidths of 39 MHz with the horizontal and vertical polarized carriers being spaced by 19.5 MHz. While the earlier satellites use a power output of 45 watts per transponder, 1F has an RF power output of 82 watts per transponder for each of the 20 active circuits. With six co-located satellites, there are currently 104 channels for television, some of which also carry multiple independent radio channels. ASTRA 1E to 1G (due for launch in 1997) will provide 56 transponders each capable of delivering a number or multiplex of compressed digital services. ASTRA 1H is due for launch during 1998 and is intended as a full in-orbit spare backup for the others in the system and an ultimate replacement for 1A at the end of its service life. During 1998 it is expected that ASTRA will obtain a second orbital slot, probably at 28.2°E.

It will be recalled that the input to a television receiver to ensure good quality pictures is in the order of 1 mV in 75 Ω or about 15 nW. By comparison, the signal level delivered to a satellite receiving antenna is about 30 dB lower at around 15 pW. Hence the receiving system must be engineered around high-gain and low-noise operation. Frequency modulation (FM) is thus used for analogue television because of its better noise and intermodulation characteristics. However, for about 14% of each TV line period, the signal actually rests at the blanking level and this would create bursts of carrier at a constant frequency that could interfere with other services. To avoid this problem, a sawtooth waveform typically at half field frequency is added to the composite video signal in order to disperse this energy across the bandwidth. This energy dispersal signal can be removed at the receiver by using a clamp circuit following demodulation. For the transmission of digital television signals, there are two forms of modulation that are likely to be important. In Europe, the format known as differential quadrature phase shift keying (DQPSK) will be used, while in the USA, a digital form of quadrature amplitude modulation (DQAM) is favoured.

THE RECEIVER SYSTEM

The receiver is designed around the basis of the multi-conversion superhet concept with the first frequency conversion being carried out within the outdoor unit that consists of a parabolic reflector antenna and a low-noise block (LNB) channel converter. The outdoor feature adds very significantly to the failure rate and faultfinding problems. The indoor part of the system follows fairly conventional double superhet processing techniques.

THE OUTDOOR UNIT

It is the function of the reflector to collect the radiated signal energy from a given satellite over a fairly wide area and then concentrate this on to a small aerial pickup circuit that may be as simple as a monopole or dipole tuned to the centre frequency of the band to be received. Since the gain of the reflector is proportional to its effective area, the first requirement is thus to ensure that it has a sufficiently large area. Too large and the antenna develops a tendency to become an aerofoil that tries to take off in a high wind. Because the reflector beamwidth is inversely proportional to the gain, then a high gain will naturally create a narrow beamwidth which increases the problems of pointing accuracy and introduces pointing errors in windy positions. The antenna size, location and fixing is thus the first problem to be solved when installing a new system. However, because of the well-behaved nature of the propagation conditions, local experience can reduce the complexity of this task.

A calculator with trigonometrical functions and the following algorithm can quickly produce the azimuth and elevation angles required for geostationary satellite reception. For the northern hemisphere, the convention is to use positive angles for satellite and ground stations west of the Greenwich Meridian (easterly locations use corresponding negative values).

1. Find the angle WEST (the longitude difference, Satellite °West − Station °West).
2. Find X = (cos WEST° x cos(Station latitude°)).
3. AZIMUTH = $\tan^{-1}$(tan WEST/sin Station latitude), add 180^0 if satellite is west of the Greenwich Meridian.
4. Calculate $Y = \sqrt{(1 + K^2 - 2KX)}$, where $K = 6.608$, the distance between the satellite and the earth centre in terms of earth radii.
5. ELEVATION = $\cos^{-1}((1 - KX)/Y) - 90°$
6. Calculate range $Z = 6378$ x Y km.

Note: for latitudes greater than 81°, $X < 0.15$ and this indicates that the satellite is below the horizon.

For a more professional approach, the software program Satmaster Pro for Windows, marketed by Swift Television Publications, has been tested. This is simple to use, requires the use of a relatively low-level personal computer system and provides a wealth of data suitable for the design of any geostationary TV system, both analogue and digital. The features provided cover antenna sizing and pointing calculations, up-/downlink power budgets for analogue and digital services, bandwidth and side lobe patterns to allow checks to be made on sources of possible interference, together with rain fade margins, tropospheric and atmospheric effects. Tables 8.1 and 8.2 are just two examples of the screens available from an extensive program suite that provides the calculations for operation with any geostationary satellite at almost any earth station location. (The authors are grateful to Derek Stephenson for the opportunity to test his suite of programs.)

Dish alignment

The provision for a fixed reflector dish is very straightforward once the look angles have been calculated. However, due allowance (about 28° to the elevation angle) must be made for those installations that employ an offset fed dish. The most important feature of the installation is to achieve a situation in which the dish is stable in high winds and is unlikely to pull itself off the structure to which it has been

Table 8.1 Downlink analysis (0.53m antenna) Produced using Satmaster Wednesday, January 29, 1997

Site name	Canterbury	
Satellite name	ASTRA 1C	

Input Parameter	**Value**	**Units**
Site latitude	51.3N	degrees
Site longitude	1.15E	degrees
Magnetic variation	3.2W	degrees
Site altitude	0.25	km
Satellite longitude	19.2E	degrees
Spot beam polarization cant	−13	degrees
Channel / transponder frequency	11.332225	GHz
Antenna efficiency	67	%
Antenna noise	68	K
LNB noise figure	1.0	dB
LNB gain	52	dB
LNB load impedance	75	Ohms
Coupling loss (waveguides and polarizers)	0.3	dB
EIRP	54	dBW
RF/IF noise bandwidth	26	MHz
FM deviation (peak-to-peak)	16	MHz
Maximum video frequency	6	MHz
Pre-emphasis / De-emphasis	2	dB
Weighting factor	11.2	dB
Demodulator threshold	8	dB
Antenna ageing, pointing and polarization loss	0.5	dB
Signal availability (average year)	99.5	%
CCIR rain climatic zone	H	
Water vapour density	12.82	g/m^3
Surface temperature	15	degrees C
Polarization	Vertical	

Satellite Look Angles	**Value**	**Units**
Elevation	28.90	degrees
True azimuth	157.34	degrees
Azimuth compass bearing	160.54	degrees
Polarization offset	−0.94	degrees

Modified Polar Mount Settings	**Value**	**Units**
Polar axis	51.97	degrees
Polar elevation	38.03	degrees
Declination offset	6.76	degrees
Apex declination	58.73	degrees
Apex elevation	31.27	degrees

bolted. Furthermore, it is important that no part of the bare metalwork should be exposed to the elements otherwise corrosion and rusting will occur.

A much more elegant arrangement for use with modern multichannel receivers is to employ a motor-driven modified polar mount antenna which will allow all the *visible* geostationary satellites to be tuned in.

Table 8.1 (Continued)

Clear Sky Result	Value	Units
Path distance to satellite	38713.80	km
Free space path loss	205.29	dB
Spreading loss	162.75	dB/m^2
Atmospheric gaseous absorption	0.16	dB
Tropospheric scintillation fading	0.09	dB
Total atmospheric losses	0.26	dB
Effective area of antenna	−8.30	dB/m^2
Antenna gain	34.24	dBi
Pre-detection noise bandwidth	74.15	dB.Hz
LNB noise temperature	75.09	K
System noise temperature	174.13	K
Power flux density (PFD)	−109.01	dBW/m^2
Total carrier power	−118.11	dBW
Total system noise power	−132.04	dBW
Carrier power at LNB output	−66.11	dBW
Signal level at LNB output (75 ohm)	72.64	dBuV
Signal level at LNB output (75 ohm)	12.64	mV
Figure of merit (G/T)	11.03	dB/K
Carrier to noise ratio (C/N)	13.93	dB
FM improvement (modulation gain)	16.65	dB
Signal to noise ratio (S/N)	43.78	dB
Picture quality (CCIR Grade)	4.3	

Degraded Sky Result	Value	Units
Signal availability (worst month)	98.44	%
Attenuation due to precipitation	0.75	dB
Noise increase due to precipitation	0.85	dB
System noise temperature	211.93	K
Downlink degradation (DND)	1.60	dB
Cross polar discrimination (XPD)	36.86	dB
Power flux density (PFD)	−109.75	dBW/m^2
Total carrier power	−118.86	dBW
Total system noise power	−131.19	dBW
Carrier power at LNB output	−66.86	dBW
Signal level at LNB output (75 ohm)	71.89	dBuV
Signal level at LNB output (75 ohm)	11.89	mV
Figure of merit (G/T)	10.18	dB/K
Carrier to noise ratio (C/N)	12.33	dB
FM improvement (modulation gain)	16.65	dB
Signal to noise ratio (S/N)	42.18	dB
Picture quality (CCIR Grade)	4.0	

The mechanism for such an antenna is shown in Figure 8.1 and the need to apply a declination angle correction arises for the following reasons. From an equatorial viewpoint, the Clarke orbit appears as a straight line passing overhead from east to west. If viewed from a high enough altitude above either of the poles, the orbit would appear to be circular. Thus, by deduction, the orbit must take up an elliptical shape when viewed from latitudes in between. Now the simple polar mount scans a circular arc as it is rotated, so that this device only coincides with the Clarke orbit in two positions. By including the

Table 8.2 Noise figures (290K) Produced using Satmaster

Noise figure (dB)	Noise temp. (K)
0.00	0.00
0.10	6.75
0.20	13.67
0.30	20.74
0.40	27.98
0.50	35.39
0.60	42.96
0.70	50.72
0.80	58.66
0.90	66.78
1.00	75.09
1.10	83.59
1.20	92.29
1.30	101.20
1.40	110.31
1.50	119.64
1.60	129.18
1.70	138.94
1.80	148.93
1.90	159.16
2.00	169.62
2.10	180.32
2.20	191.28
2.30	202.49
2.40	213.96
2.50	225.70
2.60	237.71
2.70	250.01
2.80	262.58
2.90	275.45

declination correction factor which varies with latitude, the dish can be made to scan the geostationary orbit with an accuracy of less than 0.5°.

The drive system consists of two sections, the motorized drive assembly mounted on the back of the dish and the controller unit included in the indoors unit with the main receiver. For the latter section, control may be exerted through front panel keys or via the normal remote infra-red handset. The two sections of this system are interconnected by a multiway cable. A heavy two-way pair provides the typically 35 volts DC supply to the motor. A permanent magnet DC motor is used because it is easily driven is either direction simply by reversing the supply polarity. Lighter three or four way cables are then used to carry the control signals.

Positioning control is normally obtained by some form of pulse-generating system driven by the rotating jack screw. A rotating magnet can be used to generate pulses as it passes over either a reed switch or Hall effect device. Alternatively, an opto-coupler can be used to interrupt a light beam via a shutter driven by the jack screw. In any case, these systems typically generate 20 pulses per rotation of the drive shaft. Position control is thus a matter of pulse counting and comparing these values with those stored in a semiconductor memory to provide satellite location data. Limit switches are added to stop the motor driving the dish beyond its mechanical end stops and creating

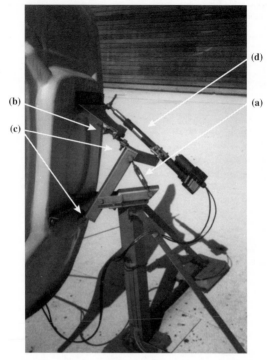

Figure 8.1 Adjustment points for modified polar mount antenna: (a) Elevated angle adjustment; (b) Declination angle adjustment; (c) Pivot axis; (d) Jackscrew and motor azimuth control.

physical damage. It is important that this mechanism is inspected periodically to ensure that the weather proofing and lubrication is satisfactory. The motor normally includes a slipping clutch to prevent the system being driven when the jack screw has seized up, but if maintenance is ignored, this safety feature may well fail. If the screw jack has to be removed for such a reason, its length should be measured before removal and the dish rotation locked off before dismantling. It is then a fairly simple matter to reset the dish mechanism back to its original settings.

The elevation angle due south and the azimuth adjustments are often referred to as the axis inclination and the east/west steering angles respectively. For installation, the declination angle is first calculated and then set on the declination adjustment. The angle of elevation for a due south heading is then obtained so that the aperture plane angle can be set. The whole structure is then rotated about its vertical axis so that with the drive mechanism set to a central position, the dish is directed towards the magnetic south bearing and then locked off. The motorized drive now swings the dish around the declination bearings. These points are all shown in Figure 8.1.

The polarizer

In the interests of frequency reuse, all four forms of signal polarization are employed to maximize the occupancy of the spectrum and extend the range of services provided. Both left and right circular polarization are used on certain satellite channels and this also helps to combat the effects of Faraday rotation of the wave energy as it passes through the ionosphere. More commonly, linear polarization, nominally both vertical and horizontal, is employed. On certain satellites, the wave polarization is offset from the vertical and horizontal by as much as 30°. Circular polarization may be converted into the linear form by inserting a block of dielectric material into the waveguide to act as a depolarizer. Ignoring Faraday rotation, only if the receiving site shares the same longitude as the satellite will the angles of polarization be coincidental. If the receiver is located to the east of the satellite, then viewed from the rear of the dish, it will be necessary to rotate the LNB in a clockwise direction to maximize the signal level and vice versa.

Early systems designed to receive both vertical and horizontal polarizations simultaneously employed two LNBs coupled to the waveguide via an ortho-mode transducer. This is a waveguide structure with two paths to feed the two LNBs but with differing path lengths so that the appropriate signals are fed to the selected LNB.

A development of this that avoided using two LNBs, involved rotating the feed probe within the waveguide through 90° using an electromechanical flip-flop that when pulsed electrically changed from one state to the other. Alternatively, the feed probe was rotated by either a stepper or servo motor. Both concepts provided good cross-polar separation but the mechanical parts were subject to wear and tear and the vagaries of the weather. Faraday rotation occurs whenever an electromagnetic wave passes through a magnetic field with the degree of rotation being proportional to the strength of the field. If the wave is constrained within a waveguide that holds a small piece of ferrite and a DC magnetic field is applied externally, the effect can be significantly enhanced. By suitable choice of ferrite dimensions and strength of magnetic field, the plane of polarization can be rotated through 90° over a fairly short length of waveguide. Typically this can be achieved with a current ranging from about ±40 mA to ±100 mA through a 100 Ω coil producing a signal loss of less than 0.3 dB. A cross-polar attenuation of more than 20 dB can easily be obtained with this system which can also be provided with a small variable current for polarization skew adjustment to maximize the signal output level.

Faults with positioning and polarization

Apart from the mechanical faults that have been mentioned, a number of electrical faults also appear to be prevalent. The control voltage circuit contains a voltage stabilizer chip that can fail, as can its associated large-value electrolytic capacitor. For systems driven by Hall effect sensors or opto-couplers any associated diodes can develop a reverse leakage that can give rise to spurious pulse counting and hence incorrect antenna pointing. In a similar way, mains-borne interference pulses caused by poor or defective mains filtering can create the problems. Failure of the voltage supply to the polarizer control may cause the loss of either vertical or horizontal or even both polarizations of signals. Failure of the insulation between the motor supply and the control lines can lead to a burn-out of the polarizer coil, therefore it is important that the connection of these cables to the system should only be made with the power off. Something that can easily be ignored when at the top of a ladder.

Low noise block converter (LNB)

The block diagram shown as Figure 8.2 indicates the essential elements of a Ku band LNB that down-converts blocks of microwave channels to the first IF. The reasons for the need of high-gain operation with very low noise for this front-end component have already been stated. From the service point of view it is therefore important that these features are maintained if sparklies on the images are to be avoided. The unit which is constructed as a replaceable integrated device consists of several low-noise amplifiers (LNAs), a low-noise converter (LNC) and multiple stages of first IF amplification. The circuit is commonly constructed from thin film micro-strip on a ceramic or similar substrate. The discrete components used are normally of the surface mounting type, for low-component loss reasons. The unit is housed in an aluminium casing that is hermetically sealed to provide weather protection. The waveguide input is sealed with a glass or plastic window, with the former material being preferred because it is much less prone to embrittlement and cracking through exposure to ultra-violet radiaton. A waveguide-type bandpass filter may be included in order to minimize image channel interference and reduced local oscillator radiation. The isolator stage also improves the rejection of image channel input, reduces local oscillator radiation and improves the impedance matching between the waveguide and the RF input stage. The multistage RF amplifiers, mixer and local oscillator use HEMT (high electron mobility transistor) GaAs devices that are selected for high-gain and low-noise characteristics. A typical single device may have a gain as high as 12 dB with a noise factor of less than 1 dB beyond 12 GHz. The local oscillator employs a dielectric resonator, the microwave equivalent of a piezo-electric crystal, for high-frequency stability. The oscillator can be tuned over a range of about 1% of centre frequency by perturbing the resonator's magnetic field with a screw which varies the air gap between the disc and the circuit casing. Monolithic microwave integrated circuits (MMICs) are progressively being introduced which reduce the size, power consumption and heat dissipation. A typical MMIC uses HEMT technology and consists of RF amplifiers, double-balanced mixer, local oscillator less the resonator and IF amplifiers. Since the LNB has to drive the indoor main receiver via a significant length of coaxial cable, it is usual to include an IF output circuit that consists of two or three stages of amplification. Since the DC power to the LNB is usually carried by the coaxial cable, it is important that these supplies are

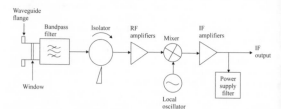

Figure 8.2   Block diagram of typical LNB.

well filtered to avoid unwanted feedback. By careful choice of components and accurate tuning of the various circuit elements, it is possible to produce an LNB with a conversion gain better than 60 dB and with a noise factor of less than 1 dB (equivalent to a noise temperature of less than 75 K).

SERVICING THE LNB

Because of the need to employ expensive microwave signal generators and spectrum analysers to ensure that any repaired LNB is restored to its makers' original specification, it is usually considered that such maintenance should only be carried out by suitably equipped workshops. However, just like a replacement with a new unit, this approach can be expensive so that the service engineer is often left either with this choice or to use his or her inherent skill and ingenuity to provide a local service. In spite of the many difficulties, provided that due consideration is paid to the following points, it is often possible to make an acceptable quality repair at an economical cost to the customer. Each LNB is hermetically sealed and weatherproofed to keep moisture and insects from the inside of the structure. Moisture can create corrosion of the waveguide and the roughness of this surface will increase its attenuation. Small insects are attracted by the warm interior and can then weave webs and the like, again to produce unwanted attenuation. The covers of the LNB are usually screwed in place and include gaskets that are not only weatherproof, but also RF-proof to constrain any electromagnetic radiation. These must be treated with care or replaced if a unit is opened for service work. It should be recognized that the removal of the covers will change the nature of the internal waveguide-like structure and influence the tuning. Also it is not easy to inject or extract signals from a microwave circuit without disturbing the normal operating conditions. Such coupling problems may be alleviated by using waveguide to coaxial cable adaptors which consist of a short section of waveguide that is blanked off at one end and contains a pick-up probe.

Signal generators from the Marconi range cover the frequency band from 1 to 18 GHz which provide both AM and FM modulation and are suitable for workshop use. A Marconi Gunn device oscillator can be used as a portable signal source and these cover the range 8 to 12 GHz with an output of 10 mW from a battery supply. A range of economically priced portable TV image/spectrum monitors and Teletext test instruments are available from PROMAX, and can be obtained from the UK agent, Alban Electronics Ltd, St Albans, Herts. AL3 6XT.

Because of the restricted bandwidth covered by each LNB (much less than 1 octave), the distortion component of a lower frequency signal generator may be used in an emergency. The third order distortion component may well be 50 dB below the normal carrier level but with an output at fundamental of 10 mW, the harmonic level would be around 0.1 µW, a level comparable with the normal RF input to the LNB.

The transistors used for LNBs are very expensive and any that are found to be faulty should be replaced by identical types. The device parameters actually form part of the circuit tuning. The first RF amplifier stage will normally be biased for a low-current/low-noise state, while the remainder will be biased for high gain. Resolder components with care; too much solder on a joint can affect its RF performance. Soldering to ceramic-based circuit boards can be difficult. The heat loss from the soldering iron bit is greater than that experienced with lower frequency boards. To avoid static electricity problems, the soldering iron bit and the user should be connected directly to a convenient earthy connection on the LNB.

Microwave circuit elements

Once the covers have been removed from the LNB, a service engineer may be presented with an unfamiliar view because the circuit elements and component bear little resemblance to those found in lower frequency systems. For example, filters, transformers and coupling devices look nothing like their LF counterparts. It is important that all screws and shields should be replaced exactly as found otherwise the tuning of the unit can be disturbed.

An open-circuit stub whose length is half a wavelength ($\lambda/2$) has a high impedance at each end, a feature that is also true for integer multiples of $\lambda/2$. Thus the half-wave stub shown in Figure 8.3(a) behaves as a high impedance to signals that make it so. At frequencies on either side the stub impedance falls, so that less signal voltage will be developed and it thus behaves as a bandpass filter. The structure shown in Figure 8.3(b) also has similar characteristics because the metallic strips are edge coupled via the electromagnetic field. By slightly varying the lengths of the strips, it is possible to extend the bandwidth of the filter. The high impedance metallic strip connected between the two low-impedance sections shown in Figure 8.4(a) behaves as an inductor if its length l is less than $\lambda/4$. The two low impedance sections have significant shunt capacitance, so that the combination behaves as an equivalent low-pass filter. By a similar reasoning, it can be shown that the circuit in Figure 8.4(b) acts as an equivalent high-pass filter.

Figure 8.5(a) shows how a $\lambda/4$ length of stripline can be used as a quarter-wave matching transformer between high-and low-impedance circuit elements, while Figure 8.5(b) shows the use of a tapered line to obtain a similar degree of matching.

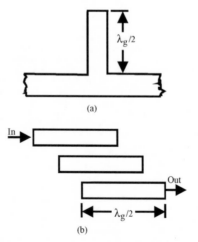

Figure 8.3 Band pass filter: (a) single open circuit stub; (b) coupled.

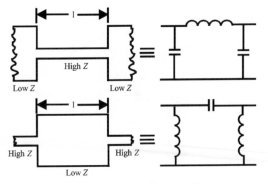

Figure 8.4 Low-pass/high-pass microwave filters.

It is often necessary to obtain signal coupling between certain circuit points and isolation between others and Figure 8.6 shows a range of such structures. When the input power is supplied to port 1 of Figure 8.6(a) the signal can arrive at port 4 by two paths. One is $\lambda/4$ long and the other $3\lambda/4$ long. The signals at port 4 are thus antiphase and self-cancelling, so that port 4 is isolated. The output power then divides equally between ports 2 and 3, and since these are separated by a strip $\lambda/4$ long there is 90° of phase difference between the signals, each of which is 3 dB below the input power. Figure 8.6(b) shows an alternative construction of the 3 dB branch line coupler. Figure 8.6(c) shows a derivation of the branch line coupler adapted for use as a selector for alternative vertical or horizontal polarization signals at the LNB input stage. The circular microwave strip is energized from the signal directed along the waveguide input and the two probes act as monopole aerials to this energy. However, there is the same $3\lambda/4$ and $\lambda/4$ split around the outer track so that signals arriving on one monopole are exactly antiphase to those arriving on the other, hence isolating the inputs to the RF amplifier that are driven from each of the monopoles.

Early LNBs designed for the FSS segment of Ku band employed a local oscillator running at 10 GHz which produced a first IF ranging from 10.7 to 11.7 GHz. As the services extended into the BSS

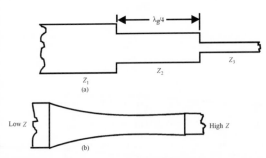

Figure 8.5 Quarter-wave transformer and impedance matching

191

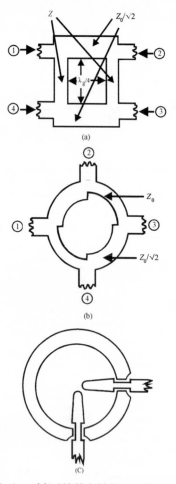

Figure 8.6 (a and b) 90° Hybrid branch line couplers; (c) Coupler adapted for selected alternative linear polarization.

segment various other combinations were employed until recently when a universal LNB was developed. This uses two switched local oscillators running at 9.75 and 10.6 GHz respectively. For low-band operation the 9.75 GHz oscillator produces a first IF range extending from 950 to 1950 MHz from the FSS segment of 10.7 to 11.7 GHz. The corresponding first IF high-band figures for the 10.6 GHz oscillator are 1100 to 210 MHz for the BSS segment of 11.7 to 12.7 GHz. The universal LNB can therefore provide for four separate channel outputs (vertical/horizontal polarity plus high/low bands) which are suitable for driving a cable head-end system.

Control of polarization and high/low-band switching

Polarity switching is usually achieved through varying the DC voltage fed to the LNB over the coaxial cable and then employing a level detector circuit to switch the appropriate RF amplifier to its microwave feed. Typically a voltage ranging from 11.5 to 14 volts and 16 to 19 volts selects the vertical and horizontal polarity respectively. A 22 kHz tone is supplied over the coaxial cable to switch local oscillator circuits. This signal has an amplitude of 0.6 ± 0.2 volts and is rectified and voltage doubled at the LNB end to provide about 1 volt DC to switch oscillator circuits. The 22 kHz tone is normally used to select the high-band mode.

FAULTFINDING WITHIN THE LNB

Within the limitations mentioned above, there are many faults that can be dealt with outside of the manufacturers' service department. The faults encountered can be subdivided into three groups: those attributable to the weather element due to electrostatic discharge (ESD – lightning, etc.) and the ingress of moisture and insects; and those that produce complete or only partial system failure made obvious by the sparkly and noisy effect produced on screen.

Outside influences

Apart from the problems mentioned earlier, moisture can leak into the IF output cable via the cable connector. In these cases, corrosion around the circuit board and connector links can produce total or partial failure, but if caught early enough a simple clean-up will often suffice. Water leakage into the coaxial cable will become apparent by a sudden increase in sparklies on the screen. Again if caught early enough, this can often be rectified by cutting off a short section of cable to remove the faulty section and replacing the line connector. In the worst case, it will be necessary to replace the whole cable run to the indoor unit. To prevent this problem occurring, it is advisable to seal the cable connector with a self-amalgamating tape at installation.

The effects of ESD are accumulative so that faults may not be immediately associated with this phenomenon. Because the LNB employs field effect amplifier devices which are susceptible to this problem, several distant lightning strikes can have the same effect as a more local one. Take care while the covers are removed, because even body static can induce the final strike.

Complete failures

When connected to a bench power supply, the current drawn with a correctly functioning unit will be in the order or 200 to 300 mA. If the current is very low or zero, this can indicate a faulty voltage regulator or an open-circuit coaxial line connector. A high current might indicate a partial short circuit through corrosion or a leaky large value power supply filter capacitor. With universal LNBs the failure of both V/H polarization and H/L band switching can cause complete or partial failures.

Partial failures

The typical LNB has three RF and three IF amplifier stages and it is unlikely that all will be simultaneously faulty. It is therefore possible to compare voltages on a high-impedance meter to obtain an indication of the activity of each stage. For the RF stages the drain voltage will typically lie in the region of $3 \pm^{1}/_{2}$ volts, while the gate voltage should be found to be around $-^{1}/_{2}$ volts. Note that the V/H switching should cause one or other of the first RF amplifiers to be turned off. The IF amplifier voltages will usually be somewhat higher than their RF counterparts but again these voltage levels should be much the same for all stages.

The local oscillator stages are difficult to check because the output is normally fairly small. This circuit may be off tune or non-functioning and either way the LNB will not work. Usually there is a very low-value resistor in series with the drain lead and a volts drop accross this will indicate whether it is drawing current or not. If the gate electrode is shunted to ground by a resistor of around 500 Ω and the voltage at the drain does not change, then the transistor is most likely to be at fault. An off-tune oscillator is difficult to diagnose with the covers off and virtually impossible with the covers on without the use of a spectrum analyser, small but noted tweaks of the tuning screw can be made to check the effects (half a turn is enough to take the signals completely out of range). This adjustment has a typical range of about 150 to 200 MHz and turning the screw clockwise increases the frequency at about 50 MHz per turn.

To ensure that noise conditions are not degraded, it is important that any transistor should only be replaced by one of the correct type.

THE INDOOR RECEIVER UNIT

There are two classes of receiver that will need to be considered, the high specification unit designed for continuous operation, cable system head end feeds and the domestic DTH receiver. Both are designed around similar principles and the block diagram shown in Figure 8.7 is intended to represent the common features of both systems. Both classes of receiver are designed to handle encrypted signals, usually by the introduction of plug-in modules. Provision is also made for the reception of all FM sound channels using subcarriers in the range 5.5–8.5 MHz.

Power for the LNB is provided from the indoor receiver via the coaxial input cable. The input stage is therefore filtered to prevent the radiation and reception of spurious signals. For reception of Ku band signals within the European region, the receiver will normally tune over the frequency range from 950 MHz to either 1750 MHz or 2150 MHz. The tuned RF amplifier stages are designed to handle input signal levels from about -60 dBm to -25 dBm (1 pW to 3 μW) without overloading.

The second IF stages operate at frequencies ranging from about 140 MHz to 480 MHz (the example shown runs at 479.5 MHz). The RF tuning and local oscillator frequencies are controlled by frequency synthesis (see Chapter 7). A remote control tuning feature can be provided via an RS232-C interface for the cable system receiver, with the control data being stored in the semiconductor memory associated with the system control desk computer.

The FM vision signal is processed using a demodulator with threshold extension to achieve a carrier-to-noise ratio (C/N) of around 5 dB. The following video signal de-emphasis time constant may be switchable to suit the particular TV signal format being processed. After filtering and automatic gain control, the video signal is split into two paths. One provides a 5.5 MHz bandwidth output, clamped to

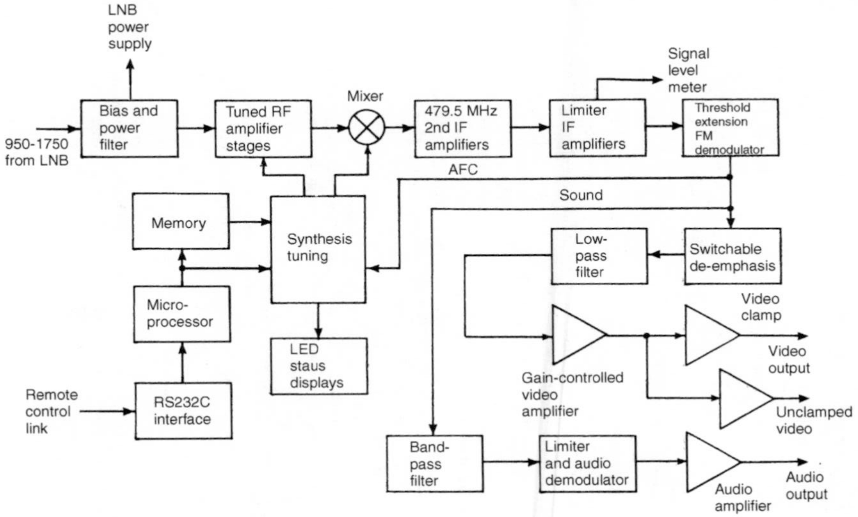

Figure 8.7 Indoor unit: receiver-decoder.

remove the energy dispersal component and the other provides a wide-band signal (8.5 MHz), unclamped, to feed any decryption decoder that may be in use.

The audio subcarrier present with the video signal is filtered and buffered before being up-converted to the standard FM IF of 10.7 MHz, this conversion being controlled via the frequency synthesizer. Demodulation then follows standard FM practice but with a range of de-emphasis time constants (50 μs, 75 μs, or J17).

Common fault conditions

Many of the common receiver faults can be located by signal tracing with an oscilloscope in the usual manner, but a good understanding of the signal flow is valuable because fault conditions in ASICs can create some unexpected signal paths. However, experience has shown that the stock fault concept is much more prevalent in these receivers than their television counterparts and the *Television* journal which collates the experience of many service engineers, is a very good resource for such guidance. The commonly occurring problems of dry joints, open-circuit or high-resistance connections tend to affect the switch-mode power supply (SMPS) section more often than the television receiver version. Probably this is due to the fact that this receiver is treated as a set top box (STB) which is thus affected not only by the heat that it generates internally, but also from the heat radiation of the TV set. Common faults are associated with diodes and bridge rectifier assemblies, dried out of high value electrolytic capacitors and small inductors used for filtering. The SCART socket which is a useful point for signal quality and level testing can also be a source of corroded high-resistance connections. Intermittent faults are usually very difficult to locate and are often temperature sensitive. The alternate application of a freezer spray and a hair drier can be used with advantage in these cases.

Smart cards and their card holder and readers used to decrypt pay channel signals can produce their special breed of problems. Many symptoms can generate the 'Card invalid' message and apart from the card edge connector contacts, there is usually a further pair of 'make' contacts that are closed when the card is removed. If these fail, they will also invoke the 'invalid' message. Contact cleaner cards are now available that carry a small section of linen cloth that is impregnated with a dry cleaning compound in place of the contacts. Inserting this card a few times will quickly clean up most of the contact problems. The cards themselves are prone to the effects of overheating, a feature that also increases the chance of contact corrosion. It is a good idea to use an old card for initial testing because some faults can actually cause a new card to become invalid. With a correctly working system, the old card should invoke the 'Your card has expired' message.

A number of strange control faults have been recorded and these are often due to corrupted data held in the microprocessor system memory. There is usually a switch included in the reset circuit to clear the memory completely to allow reprogramming. With so many channels now available retuning becomes a daunting task and Table 8.3, which is an extract from the listings of PACE Link, gives an example of why this useful first-line approach is often avoided. However, there are now some useful aids available to the engineer which allows the receiver tuning data to be downloaded into a personal computer (PC). The PACE Link system has been developed to enable the receiver tuning to be edited via the keyboard using an interface coupled between the decoder/scrambler SCART socket and the PC serial port. The software provides a wide range of facilities specifically for PACE Ltd

Table 8.3 PACE MSS 508-IP

No.	Name	Freq.	Pol.	LNB	Aud.	Sat.
59	ASTRA 59	10.862	Hor	1	StA	1
60	ASTRA 60	10.877	Vert	1	StA	1
61	ASTRA 61	10.891	Hor	1	StA	1
62	ASTRA 62	10.906	Vert	1	StA	1
–	– –	–	–	–	–	–
69	ASTRA 69	10.700	Vert	1	StA	1
70	DSF	11.523	Hor	1	StA	1
71	VOX	11.273	Hor	1	StA	1
72	S3	11.186	Vert	1	StA	1
–	–	–	–	–	–	–
78	GALAVISION	11.127	Vert	1	StA	1
79	CINEMANIA	11.656	Vert	1	StA	1
80	DOCUMANIA	11686	Vert	1	StA	1
–	–	–	–	–	–	–
85	BBC R3	10.979	Vert	1	StC	1
86	BBC R4	11.553	Hor	1	Mo5	1
–	–	–	–	–	–	–
91	SKYRADIO	11.318	Vert	1	StB	1
92	SUPER GOLD	11.171	Hor	1	Mo4	1
93	IRISHSAT	11.538	Vert	1	Mo7	1
94	RTE	11.538	Vert	1	Mo5	1

receivers. The features include reading the current receiver tuning data, editing any entry, interchanging and updating channel settings, uploading the new set-up into the receiver, together with preview and printout of the data tables. These features are also particularly useful at the installation stage of a new receiver. Before clearing the memory of any suspect system, it is only necessary to download the current tuning data, clear the memory, edit the data table and reload the new channel information. At least at this stage, the engineer knows that the memory is not corrupted. The system described above was loaned by Gordon McCrae, and Pace Link (UK) Ltd, Kesh, Co. Fermanagh, BT93 1TF for which the authors are indebted. Other similar designs are available for other makes of receiver and a number of simpler EEPROM-based circuits have been described in *Television*.

THE DIGITAL TV RECEIVER

From 1998 onwards there will be a significant shift towards the digital transmission of television throughout the world. Initially, the promised new services which will be compatible with both 4:3 and 16:9 aspect ratios will be delivered via a set top box (STB) to ensure compatibility with current receivers. Although much of the receiver signal processing will be common to satellite, cable and terrestrial delivery, the decryption system may vary according to the programme source. This concept will simplify the problems of creating an integrated decoder/receiver which at a later date will ultimately replace the STB. Figure 8.8 explains the high degree of commonality of signal processing for the three delivery methods. While this shows a large number of individual blocks, the current state of the art can reduce this to about eight VLSI ASICs. Within about two years it is anticipated that the technology will provide

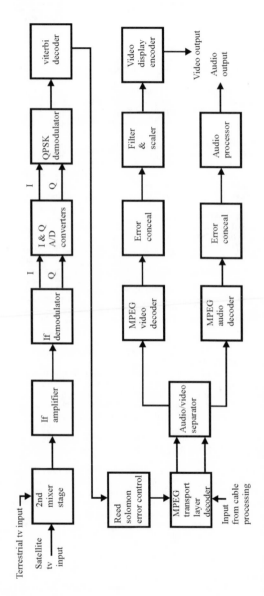

Figure 8.8 Block diagram of set top box for digital TV.

the same functions in about four devices. Because cable delivery generally provides a higher S/N ratio than either radio broadcast methods, there will be provision for only one layer of forward error control.

Both the terrestrial and satellite delivered signals are input to a second mixer stage in the manner of the current satellite receiver and the output from this is then processed via gain-controlled IF amplifiers and a suitable analogue demodulator. Following demodulation, the signal is converted into digital format before digital demodulation. At this point, there are differences between the European and American receivers. The multicarrier coded orthogonal frequency division multiplex (COFDM) is favoured in Europe, while the Americans prefer to use a version of multilevel quadrature amplitude modulation (QAM). At the error control stage, the decoding of the Reed-Solomon format includes Viterbi decoding. This is a maximum likelihood method in which any errored bytes are replaced by those which possess the least number of errors. It is chiefly this stage of error control that is omitted from the cable delivery method. After descrambling the MPEG data packets, the signal is split into sound and video channels for further decoding and error control. Error concealment works by modifying any errored bytes in one of the following ways: ignore the value and replace with a zero, interpolate between two adjacent unerrored bytes or repeat the last unerrored value. The video output can then be encoded into either CVBS, S-Video or RGB, while the audio component can be encoded for either MUSICAM or Dolby AC-3 systems.

Since the reliability and servicing problems are at present very much unknown quantities and it is anticipated that the system will be microprocessor controlled, the concept of self-diagnostics can be usefully employed. It might be worth pointing out here that the signal is in analogue format right up to the inputs of the A/D converter stages.

CHAPTER 9

VIDEO RECORDERS AND CAMERAS

Warning note. The use of the correct service manual for each machine can be critical. Although many badged machines appear to be identical, some of the signal processing circuits can be very different. Even within the same nominal type and series, VCRs and cameras can show similar variations.

All the current magnetic recording systems are derived from the basic principle of transferring the magnetic effects of an AC current that is generated from the signal to be recorded, into a corresponding permanent magnetic effect on a layer of magnetic particles deposited on a plastic base. The relationship between the cause and effect; the recording current and the recorded signal is non-linear. Thus the major aim of these systems is to overcome this particular problem. In general, this is achieved by biasing the recording signal current so that the action occurs in the almost linear region of the magnetic material's transfer characteristic. For both audio and video applications, the bias signal current is provided from an AC source and the two currents are added together in the coil of the magnetic record head. The replay system is virtually complementary in that the changing magnetic flux on the recording medium creates a corresponding signal voltage in the record/replay head mechanism.

The overall frequency response largely depends upon the relative head-to-tape velocity and the width of the gap in the record/replay head which for video is typically in the order of about 5 m/s and 0.3 μm respectively. The characteristics of the recording medium are restricted to a range of about 10 octaves which is adequate for audio applications with a range of about 20 Hz to 20 kHz. The corresponding frequency range for video signals extends over rather more than 18 octaves, a problem that is countered by modulating the video signal on to a carrier frequency. Since FM in general provides a better S/N ratio than AM, this form is chosen for magnetic recording. The video signal is split into its two components, luminance and chrominance, and the luma signal with a bandwidth of about 3.5 MHz used to frequency modulate a carrier of nominally 4.3 MHz so that the carrier swings from about 3.8 MHz to 4.8 MHz between sync pulse tip and peak white. The FM modulation index is thus typically about 0.3. Because this FM signal is of constant amplitude, it is used as the bias signal for the chroma component. However, since the original chroma signal is modulated on to a 4.433 MHz carrier, this component lies in the noisy region of the system transfer characteristic. The chroma signal is therefore down-converted on to a colour-under subcarrier frequency for the recording process.

To increase the effective head-to-tape speed and at the same time maximize the use of the tape surface area, each field of video signal is laid down in very narrow strips of about 50 μm wide, at an angle of about 4° to 5° to the tape edge. The head drum rotation is such that the tape and heads move in the same direction to minimize the tape-to-head friction and extend the lifetime of both components.

BASIC SIGNAL PROCESSING

The aerial input signal is simultaneously applied to a UHF tuner unit and a wideband loop-through amplifier to provide for direct TV reception as shown in Figure 9.1. The down-converted IF signals from the tuner are amplified under AGC control, then demodulated and separated into the four components, sound, luma, chroma and sync. The FM sound carrier is demodulated, amplified and equalized before being added to the bias oscillator signal for recording in a narrow track along the top edge of the tape. The bias oscillator which runs at around 60 kHz is only operative during record, and drives a full width erase head to wipe out all previously recorded video, audio and control signals. For stereo operation, the audio track is divided into two by a very narrow guard band and to ensure compatibility, the mono machine replays these in parallel from a single replay head. An alternative high-quality (HQ) audio is in use that employs depth modulation of the recording tape. In this case, the audio signal is frequency modulated on the carrier at about 1.2 MHz and then fed to a pair of audio record heads on the drum assembly. These lead their respective video heads and are positioned at a height so that the audio signal is recorded in the centre of the video signal track. Since the peaks of these constant amplitude and relatively lower frequencies are present for a longer period of time, these signals penetrate deeper in the oxide layer where they can be easily recovered on playback, at high quality and without mutual interference between the video and audio signals.

The luma and chroma signals are processed separately before being applied to the rotary record heads through suitable wideband amplifiers and a rotary transformer.

The field and line sync pulses are used to synchronize the head drum and capstan motor servo systems. The field sync pulses are also used to record a control track signal along the lower edge of the tape to provide part of the synchronism mechanism during replay. The head drum and capstan DC motors are driven through motor drive amplifiers (MDA) that are controlled in speed and torque via system feedback signals.

On replay, the recorded signal components are recovered from the tape via the head system, the signals are again processed separately before being encoded by the UHF modulator to provide the aerial output signal. This stage ensures that the new TV signal lies in the range between channels 32 and 40, with the actual value being chosen to minimize the effects of adjacent and co-channel interference from the direct broadcast services.

TAPE LOADING MECHANISMS

For video cassette systems there are a number of variations in the cassette loading mechanisms, but in general these fall into one of two categories. The method employed in the VHS system relies on two poles that pick up the tape in two positions and then with a lazy tong action, pull the tape out of the cassette and wrap it around the head drum in the manner shown in Figure 9.2. (This is sometimes referred to as M loading.) The alternative, which is sometimes referred to as U loading and is employed with the Beta system, picks up the tape at one point and then with a rotary action wraps it around the drum to achieve a similar effect. In either case the wrapped and loaded tape makes contact with the drum for rather more than 180° to form an omega-shaped loop. To ensure correct operation, there are a number of sensors placed in strategic positions as indicated in Figure 9.2 and

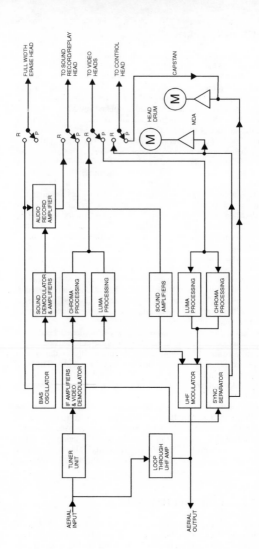

Figure 9.1 Signal processing stages for VCR machine.

202

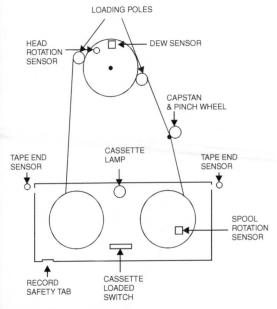

Figure 9.2 System control sensor locations.

these, together with user inputs, control the correct system operation.

THE TAPE PATH

Figure 9.3 provides a generalized view of the tape path of a typical VCR machine with the individual components arranged in-line to better explain their effects. The tape is transferred from the feed spool across the various components of the mechanism and on the take-up spool. In the interests of good record/replay quality it is necessary to maintain a constant head-to-tape contact, tape tension and smooth movement. A significant part of this section of the VCR is concerned with damping out any such fluctuations. Until the cassette is loaded into the machine, the reel/spool spindles are locked by ratchets to prevent tape spillage. The lacing-up operation is usually carried under the control of a small DC permanent magnet-type motor and once the loading operation has started, the cassette is locked into place. Reading from the feed spool end of the mechanism, the tape first passes over a spring-loaded lever designed to act as a tape slack/tension sensor which is also linked to the spool braking system to ensure a constant tape tension. The first guide feeds the tape across the full width erase head. This is followed by an impedance/damping roller that is used in the VHS system to damp out any excessive tape fluctuations. Further guides and slant poles ensure that the tape passes accurately and consistently on to and off the head drum while the heads are scanning the tape. The VHS system has a further accurate tape path guide in the machined edge on the lower drum surface. The

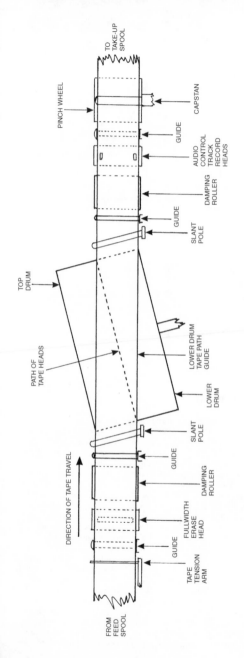

Figure 9.3 Notional tape path components.

Beta system utilizes a pair of small spring-loaded plungers on the top drum to achieve the same end. After passing the rotary heads, the tape is guided over the audio/control track record/replay head and on to the capstan where it is trapped between the shaft and a neoprene-faced roller. It is the capstan that creates the controlled tape speed but the take-up and back tension is developed by the take-up drive motor, the feed spool brake or drive motor. This latter motor can be lightly energized in reverse to provide the reverse tension. Slipping clutches are included on the reel drives to prevent tape damage.

A significant number of damage problems result from this part of the system. A common fault arises from dampness which causes the tape to stick to the head drum when the oxide layer tends to be stripped off. Dew sensors are thus incorporated to prevent the machine from starting up under this condition. Faulty drives, take-up clutch or braking mechanism can cause the tape to spill over into the revolving mechanism and become damaged. Tape guides and poles can become loosened, distorted or bent through maltreatment, as can their mounting plates. Such misalignment can cause tape crinkling at the edges to create either audio or control track problems. Furthermore, guide pole misalignment can create noise bars at the top or bottom of the picture. If the noise bar occurs at the top, the pole on the feed side could be misaligned and vice versa. Most of the sensors shown in Figure 9.2 are designed as part of the control for this area of the machine.

VCRs rely extensively on microprocessor control and these devices can create some very interesting and thorny faults because of a corrupted memory. Often it will be found advantageous to clear the memory using the service all-clear button and then reprogramme the control system anew.

PROCESSING THE LUMINANCE SIGNAL

The demodulated video signal is low-pass filtered to cut off above about 3.5 MHz to ensure that any vestige of chrominance information is removed. (See Figure 9.4). The luma signal is then amplified under AGC control, maintaining the 3:7 amplitude ratio between sync pulse and peak white. Pre-emphasis is then employed to boost the HF signal component and improve the S/N ratio. Since this action is akin to differentiation which creates over- and undershoots on sudden signal transitions, a DC clamp circuit ensures that the video signal is clamped to black level during each line blanking interval, while the white clip circuit removes the overshoots. The luma signal is then applied to the FM modulator stage. Overdeviation is in general prevented by the adjustment of the black level/sync pulse tip clamp and the AGC operating point. The signal is then amplitude limited to remove any unwanted AM component that might create false colour on replay. The luma signal is further amplified and equalized before being applied to the record heads via a rotary transformer. Because the signal at this

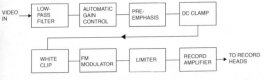

Figure 9.4 Basic luma processing for record.

point will act as the bias component for the chroma signal, it is important not to under or over drive the writing current which must be set to such a level as to provide a good S/N ratio during replay. The signals to each of the two record heads are switched alternately at the head driving stages via a bistable circuit driven from the field sync pulse. In addition to this field rate switching, it is also necessary to include record and replay switches to cater for this dual purpose function. In order to minimize cross-talk between adjacent video signal tracks, each head has opposite slant azimuth gaps, typically either ±6° or ±15° depending on the system.

The luma replay follows almost complementary processing with the signal being developed across the same heads and rotary transformer. Because the recovered signal has a relatively low amplitude, it is important that the first stage should include a low-noise amplifier. After field rate switching and amplification, the luma and chroma components can be separated for separate processing with low- and high-pass filters that have cut-off frequencies around 1 MHz. Some simple arithmetic will show that each of the 312½ lines of video signal occupies rather less than ½ mm of tape track. Therefore imperfect tape-to-head contacts can result in sudden loss of signals that are referred to as *drop-outs* which appear as blank spaces on the screen. Drop-out compensation is applied before demodulation by recycling each line of video through a delay line. Because the video signal changes only relatively slowly in the vertical sense, when a drop-out is detected, the video output is briefly switched to the previous line to mask the signal loss. The video signal is then amplitude limited, demodulated and de-emphasized in the normal FM manner. Any luminance cross-talk that was not removed by the slant azimuth head gaps can be removed by taking the current and delayed previous lines and then adding and subtracting them to cancel the unwanted component.

Because the luminance signal was band limited to about 3.5 MHz, the loss of HF is disguised by the addition of aperture correction or crispening circuits. These have a differentiating effect to boost the HF response and can give rise to overshoots which need to be clipped. Because the action of this circuit can accentuate the noise and sparkly effect, it is usually clamped to be inoperative below some presettable signal level.

PROCESSING OF CHROMINANCE SIGNAL

It will be recalled that the chroma signal represents a phase- and amplitude-modulated signal superimposed on a carrier frequency running at nominally 3.58 MHz (NTSC) or 4.433 MHz (PAL) and in both cases the bandwidth is about 2.5 MHz. However, since the human eye is very much less sensitive to colour than brightness information, this bandwidth can be restricted to about 1 MHz with relatively little loss of signal quality. All of the analogue VCR systems remodulate the composite chroma signal on to a lower subcarrier frequency (colour-under frequency) that restricts this bandwidth to about 1.2 MHz. The principle involved is the same for all standards and only the actual frequencies vary. Figure 9.5 is representative of the method and this uses a crystal-controlled oscillator that operates in both the record and replay modes. Taking the VHS system as an example, the crystal oscillator runs at 4.4355717 MHz and provides one input to Mixer 2, while the other input is derived from a phase locked loop (PLL) oscillator that is locked to the off-air line frequency and runs at $40 \times 15.625 = 625$ kHz. The output difference frequency after filtering is exactly

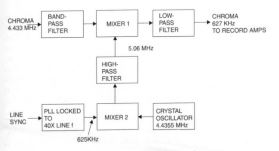

Figure 9.5 Chroma processing-record mode.

5.0605717 MHz and when this is mixed in Mixer 1 translates the
4.43361875 MHz chroma signal down to 626.955 kHz (or 40 × 15.625
+ 1.53 kHz). In this way, the recorded line frequency has been locked
to the off-air standard. During replay, the same crystal oscillator
frequency beats with the accurately controlled line frequency to regenerate
the chroma signal at the original 4.43361875 MHz with
virtually the same standard of accuracy. Because there is no provision
for guard bands between adjacent video tracks, the slant azimuth heads
are not entirely effective in cancelling out the lower frequency chroma
cross-talk. Each format uses a similar electronic approach to provide
cancellation. As described earlier, the video signals are laid down across
the tape width in parallel and with the line pulses aligned. Thus there
is a fixed phase relationship between the wanted signal and the cross-
talk component. The VHS system records from head A with the direct
version of the colour-under signal but from head B with signal that is
phase retarded by 90° on a line-by-line basis. During replay, head A
recovers its own signal plus a small degree of cross-talk from adjacent
tracks which was laid down by head B. The replay signal from head B
is phase advanced by 90° again on a line-by-line basis. The cross-talk
component is then cancelled by introducing a two line delay and add
circuit. The Beta system uses two colour-under subcarriers derived
from either 43.875 or 44.125 times line frequency to produce the same
effect. This results in a phase shift of ±45° on adjacent heads to
produce the same effect. The Video 8 system uses a system more closely
related to VHS with head 1 recording a phase continuous chroma
signal on a subcarrier of 46.875 times line frequency alongside the
head 2 signal with a 90° line-by-line phase-retarded component.

ANCILLARY LUMA/CHROMA PROCESSING AND
FAULT CONDITIONS

Automatic colour control (ACC) is necessary in both record and
replay modes to maintain the luma/chroma gain balance and
compensate for any signal level fluctuations. On record, this gain is
under the control of the colour burst amplitude. Colour killer
circuits are also incorporated in order to shut down the chroma
channel on monochrome signals to avoid false coloration. Again,
as in the colour TV receiver, it is necessary to compensate for the
different propagation delays through the luma and chroma amplifier channels. Hence a small signal delay is incorporated in the luma
channel.

In order to be able to monitor a recording as it proceeds, the video and audio baseband signals are modulated on to a locally generated carrier, nominally on channel 35 or 36 to provide an input at UHF to a standard TV receiver. Usually the signals are tapped off late in the record chain and just ahead of the feed to the record amplifier stages. They are then cross-coupled by switching to a similar corresponding point in the replay chain. This electronics-to-electronics (E – E) mode of operation simply bypasses the magnetic stages of processing. The feature can be usefully employed in faultfinding for complex problems. By making a direct recording on the suspect machine and then playing this back and comparing it with a replayed standard test tape in the E – E mode considerably helps to narrow down the problem area. Again, it is common to find baseband and composite signal components available at the SCART connector which also provides another source of signal quality examination.

A standard test tape can be a valuable tool for faultfinding, but these are expensive and can easily be ruined. Therefore although a locally produced tape might be of lower quality, this can be used with advantage. Such a tape should include a section of standard colour bars, plus any other patterns that might be of use for picture alignment purposes, together with a blank section on which a replay from the machine in question can be made and compared. The tape might also usefully include a range of audio test frequencies such as 1 kHz, 4 kHz and 8 kHz. Incorrect record current for the FM luma component will be seen as a noisy picture if too low and compressed saturation primary colours if too high. If a spaghetti type of interference is noted, then this might be due to a high luma record current or poor limiting in the FM channel, while misadjusted FM deviation controls can give rise to some very strange effects. Excessive amplitude replay chroma signal can give rise to patterning particularly on vertical edges of an image. Unlocked and generally spurious coloration might well be caused by an unlocked chroma colour-under conversion circuit. Difficulties can be introduced by the false operation of the colour killer circuit which can incorrectly shut down the chroma channel. Incorrectly set pre-emphasis and white/dark clip circuits can introduce sparklies and degraded the sharpness on vertical edges of an image.

SERVO CONTROL SYSTEMS

The tape drive mechanism is commonly managed by the use of four DC motors, two to drive the tape reels, one to drive the linear tape movement via the capstan and one to rotate the head drum. DC motors are preferred because they are simple to control for torque, speed and direction of rotation. The reel drive motors are designed to transport the tape in both forward and reverse direction for normal record/replay or to produce a fast winding action. In order to maintain a constant head-to-tape contact and tape tension as the tape winds from one spool to the other, the rewind motor can be lightly energized in reverse and this action is controlled by the tape tension sensor via a feedback loop.

For a 625 line system, each record head must lay down $312\frac{1}{2}$ lines of video signal during every half revolution and then switch over during the VBI to the second head to record the following field. The drum motor must therefore rotate at precisely 1 revolution in every 40 ms, or at a rate of 1500 rpm, with a head switching rate of 25 Hz. The synchronism for this control is obtained from the field rate sync pulses that are recorded along one tape edge. Since linear tape movement is controlled by the capstan drive, it is this motor that is effectively responsible for the recorded field rate sync pulses. Therefore line or frame slip is caused respectively by either the drum or capstan motors running at an incorrect speed.

Figure 9.6(a) shows the basic principles of a servo motor speed control system. An accurate crystal-controlled oscillator creates a divided down gating signal that is used to control a sample and hold circuit. A tacho-generator attached to some rotating mechanism that is driven by the motor generates a pulse either from a rotating magnet over a pick-up coil or Hall effect device, or via a shuttered opto-coupler. This pulse is then shaped into a ramp waveform which is then sampled to provide a DC voltage proportional to the relative timing of the ramp and gating pulse. As indicated by Figure 9.6(b), as the timing delay varies, so does the amplitude of the DC output voltage. This voltage is then used to control the motor speed via the output current of the motor drive amplifier (MDA). Fault conditions in these circuits range from loss of tacho pulse which causes the servo to free run at an incorrect speed, a change of time constant in the sample and hold circuit which can give rise to jitter, or loss of reference frequency. In general, all of these conditions can be monitored and identified using an oscilloscope coupled to the sample and hold circuit.

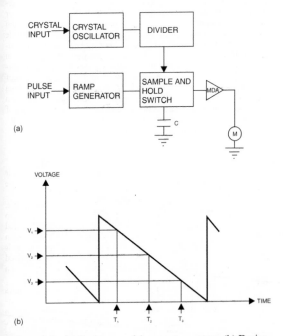

Figure 9.6 (a) Basic servo-driven motor system; (b) Derivation of control voltage.

Digital control of servo loop

This system uses two counter circuits to control the servo system. A crystal-controlled oscillator generates an accurate clock pulse stream at a fixed frequency. The arrival of the reference pulse starts counter 1 and the incrementing value is loaded into a memory register. At the instant of the arrival of the sampling pulse, the count ceases and the value held in the register represents the time difference between the two pulses. During this period, the second counter has been totalling the pulse stream direct from the clock circuit. The two counts are then compared and any difference is converted into analogue format to be used as DC control voltage as in an analogue control circuit.

THE BRUSHLESS DC MOTOR

Some of the virtues of the DC motor have already been pointed out; however, the problems associated with the commutating brush gear of the basic motor can be resolved by using magnetically controlled semiconductor switches that produce the necessary current reversals. The motor then becomes more reliable, produces less interference and runs more smoothly. Figure 9.7 shows one particular configuration for such a motor. The two windings W_1 and W_2 are fixed, while the rotating part is formed by a permanent ring magnet, magnetized with three pole pairs. Two Hall generators Hg_1 and Hg_2 are mounted at right angles to each other and spaced about 1 mm from the periphery of the rotor. When a magnetic pole passes each generator, a pulse of about 0.5 volt amplitude is produced of positive or negative polarity. These pulses are then used to operate semiconductor switches within the controller IC to produce the necessary commutation of the supply current. If the switches are operated in the sequence a, b, c, d, a . . . , the motor runs clockwise. By changing the polarity of a control line in the IC (not shown), this sequence changes to d, c, b, a, d . . . , and the motor reverses.

DC power is provided via a direct-coupled motor drive amplifier. Since the winding currents all flow to earth through the resistor R_1, the voltage developed across this can be used to provide the feedback signal to drive a constant speed controller loop.

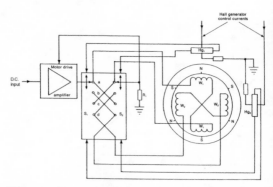

Figure 9.7 Basic principles of Hall effect motor control system.

Faultfinding in this type of motor is more complex than with the basic type of machine. Not only may faults develop in the motor windings, but in the control IC and the motor drive stages, the Hall device bias current can also fail. Since these are often provided via regulator stages, these add another source of faults. If the supply voltage is low the motor will take longer than usual to run up to operating speed and then fail to develop the necessary driving torque. A poorly filtered power supply can lead to jitter in the motor rotation.

SUPER VHS AND 8 MM SYSTEMS

The search for improved image quality with lower noise level, better linearity and less distortion has spawned these new formats. The availability of a new tape with smaller oxide particles not only produces a smoother surface with lower noise, it also provides for higher coercivity and remanence. When combined with a new head material that allows a reduction in the width of the head gap, the S-VHS machine produces almost broadcast quality images. In order to maintain compatibility with earlier recordings, the parameters of the mechanical parts of the new system remain unchanged. The S-VHS cassette carries an additional ident hole to enable the recorder to switch automatically between standards. The FM luminance parameters have been changed so the sync pulse tip corresponds to a frequency of 5.4 MHz with peak white at 7 MHz. The same colour under frequency is retained for the chroma signal. A narrower head gap allows for a higher frequency response even though the video track width has been marginally reduced. Provision is made for direct inputs and outputs at YC (luminance and chrominance) to reduce the degree of processing in certain applications. Provision is also made for two audio systems. A conventional two-track linear mono or stereo system with a frequency response extending from 40 Hz to 12 kHz includes Dolby noise processing. An FM stereo depth modulation system covering the frequency range from 20 Hz to 20 kHz with a dynamic range of better than 87 dB employs carriers of 1.4 and 1.8 MHz. The azimuth angles of the video and audio heads are as in the standard VHS system. Using this technique, the audio signal cannot be edited without re-recording the accompanying video signal.

The 8 mm systems use a tape that is about 33% narrower than that for VHS so that this cassette is smaller. These systems employ an automatic track following (ATF) technique that was originally pioneered in the Philips V2000 system. This allows the heads to move laterally across the tape width in order to precisely follow the video tracks and is achieved by laying down a specific signal pattern during record so that on replay the heads follow exactly the same path. The audio signal occupies a track along the tape edge and with the stereo signal being recorded as digital PCM. The colour-under frequency for the chroma signal is placed at about 732 kHz and the FM luminance frequencies are 4.2 MHz for sync pulse tip and 5.4 MHz for peak white. With a head gap width of about 0.26 μm, a resolution of about 260 lines can be achieved. The Hi8 system uses the same mechanical features but with a head gap width of 0.2 μm and FM luminance frequencies of 5.7 and 7.7 MHz to achieve a resolution of about 380 lines. A degree of compatibility exists between the systems but the standard Video 8 machine cannot replay Hi8 recordings. As one would expect, drop-outs and overmodulation effects first tend to become noticeable on the high-band system.

POWER SUPPLIES

These are usually based on the switch-mode principle that operates in the same manner as those used in the television receiver, but not locked to line timebase frequency. The individual lines range from 5 volts up to about 30 volts, with both positive and negative polarities but not all need to be stabilized supplies. Voltage regulators are therefore often a common cause for concern. Because of the extensive nature of the switching requirements in the VCR, many of the functions are managed via diode or transistor switches, often controlled via a microprocessor. Dry joints and electrolytic capacitors that dry up or become leaky produce many of the common faults. Because of the feedback within the SMPS a loss of capacitance which produces a drop in output voltage can cause the system to increase the output to meet this new demand. It has not been unknown for this action to burn out capstan or drum motors and even burn the PCB if the condition exists for too long. The power supply section of a VCR is often coupled to the signal boards via a number of multiway rainbow cables. The connectors of these are often treated with a strip of glue to prevent unwanted disconnections during normal operations. It is not known for such glue to become oxidized and conductive with time and temperature, thus creating partial short circuits.

MECHANICAL PROBLEMS

With the passage of time, varying temperature, humidity and the continual transportation of magnetic tape across the mechanical part of the VCR, there is bound to be a considerable build-up of dust, tape particles and other unwanted debris within the casing of the machine. Due to friction and electrostatic forces, much of this will adhere to the metallic parts of the system. Head cleaning tapes that consist of a band of lint-free fabric impregnated with a proprietary cleaner that can be run through the machine in place of the normal tape are often the user's front-line approach to degrading replay signal-to-noise ratio. Sometimes these are effective but if used too often, they invariably create more damage to the tape heads because they become impregnated with abrasive material. Dismantling and the careful cleaning of the tape path with isopropyl alcohol and lint-free cotton buds can do much more for noisy images, particularly if this is caught in the early stages. At the same time the lubrication of the moving mechanical parts should be checked, but care should be taken to avoid over-lubrication as this can create even more problems. Wear of the video heads shows up as noisy replayed images, although much of the noise may have got in through the record channel. Before changing heads though, it is as well to check the earthing brush at the centre of the head drum for good continuity, because this can create a similar fault.

Defective pinch rollers and tape guides can cause the tape to ride up or down the capstan and bear heavily on the tape edge to cause crinkling and hence noise. Excessive take-up torque or weak back tension can give rise to similar problems; while faulty braking can confuse the control system into various faults of non-co-operation. A dental mirror can be usefully employed to check the state of many of the hidden parts of the mechanical control section. A standard test tape is necessary to obtain optimum adjustment of the tape guides and general alignment. Many of the critical mechanical adjustments require the use of jigs, tools, tension gauges and even a dial gauge to check the concentricity of the video drum. The precise adjustments tend to vary from maker to maker and therefore it is important to follow the processes laid down in the service manual. Furthermore, after extensive mechanical adjustments, many of the electronic circuits need to be optimized and this needs the services of an oscilloscope with a sensitivity of about 1 mV and bandwidth of at least 25 MHz.

AUTOMATIC VCR PROGRAMMING

The VCR timer which is microprocessor driven is designed to be programmed via the user's remote control handset and the process varies from machine to machine so that it is not particularly user friendly. While in standby mode, the system detects the programmed start time and automatically starts to record. At the end of the period, the system switches the machine back to the standby mode to await the next phase of operation. The VideoPlus system provides an extension of this facility. Each programme in the listings is allocated a number consisting of up to seven digits that identify every combination of date, start time, duration and channel number. The user then simply enters these digits into the system memory via the handset. The disadvantage of Video-Plus lies in the fact that it cannot detect programmes that are run at different times to those advertised. If a programme runs late because of an over-run in the previous programme, the recording is simply cut off short.

Programme delivery control (PDC) or video programming system (VPS) are systems that have been developed to operate through the broadcast teletext system and so avoid the main problem of Video-Plus. Both systems are compatible with each other and the Country, Network Identification (CNI) system. They are configured in highly integrated chip sets that allow the decoders to respond automatically to either system. Each programme listed in the teletext television guide can be selected by the user under cursor control from the remote handset. Selection then downloads the date, channel number and start and stop times into the system memory. Each television programme carries a start and stop time data tag so that the decoder simply scans until it finds a match and then switches the VCR into the appropriate mode to provide accurately timed recordings. The timing is based on the Unified Date and Time (UDT) system and if UDT is not being transmitted, the system locks to the real-time clock transmitted in the header row of each teletext page. The PDC system uses standard teletext coding and bit rate with the data contained in Packet 8/30. The VPS information is included in the vertical blanking interval on Line 16 of each television frame and the data is transmitted at 2.5 Mbit/s using Manchester bi-phase coding. Fifteen, 8-bit words are transmitted at a repetition rate of 40 ms at the 50 Hz field rate.

CHANNEL 5 TV (C5B)

Originally channels 35 to 37 were seen in the UK as a break between the UHF Bands 4 and 5 and thus became useful for the remodulation systems such as VCR and satellite systems. Channel 38 was also avoided because of the needs of radio astronomy. However, France and Ireland, among other nations, have used these frequencies for television broadcasting. The introduction of Channel 5 television using mostly channels 35 to 38, plus others within the major UHF sector, has involved a significant new band plan for the UK. In particular, many of the VCR installations within the new region have had to be retuned to a new intermediate carrier frequency. In spite of the large population density, much of the south-east of England has been precluded from this new service due to the possibility of mutual interference.

Retuning of the affected systems basically involve using the input of a captioned test pattern to simulate the new broadcast station at, say, channel 36 and then adjusting the RF output from the VCR/satellite tuner to a clear spot in the UHF spectrum where there is no mutual interference, followed by adjusting one spare TV button to the C5B frequency and the AV button to the newly set VCR/satellite output frequency. In practice this operation is not quite so easy. An alternative approach that is somewhat more expensive has been suggested that involves using a set top box consisting of a remodulator to shift the C5B frequency right up to the little-used channel 69. Normally the atmospheric temperature decreases with altitude, but under high pressure and dry conditions, an inversion (advection) can occur that creates extended propagation conditions for VHF and UHF signals. This gives rise to adjacent and co-channel interference with television that can last for several days during the summer months. Thus it is quite likely that C5B signals will create problems for systems in south-east England that retain the older tuning arrangements.

CAMERAS

Camera repairs are really a task for specialists. To achieve the high standard of repair, it is necessary to equip a workshop with at least a wideband oscilloscope/vectorscope, standard colour filters and test cards, plus a range of special test lead extensions. Having made this point, useful first line approach work is possible and a subscription to *Television* can turn out to be a valuable resource.

Basically, a video camera consists of a record/replay machine that provides its own signal input via a pick-up sensor, commonly a charged coupled device (CCD) and a liquid crystal display (LCD) monitor that doubles as an electronic view finder (EVF). The system is further complicated by the addition of two small DC motors to drive the zoom and auto focus lens. The latter adding further complexity due to the inclusion of a sensor and a closed loop servo system. The cameras made for the domestic entertainment market represent a significant development in miniaturization together with a high level of circuit integration. As an example of this miniaturization, the compact VHS-C machine uses a smaller cassette and head drum. In order to maintain compatibility with the full-sized VCR, the drum has been reduced from 62 mm to 41 mm diameter and includes four heads, its speed increased from 1500 rpm to 2250 rpm and the tape wrap increased from 180° to 270°. The CCD image sensors have a number of significant advantages over the earlier cathode ray tube pick-up device. Because of the lower operating voltages they have a much lower power consumption, have a longer lifetime and need fewer adjustments for setting up and for temperature and ageing effects. Multiple horizontal or vertical lines usually indicate a faulty device and under these circumstances the complete CCD unit should be changed. Faults associated with the CCD unit usually relate to the power supply, the clocking network or even the input/output lines.

Although these cameras are light and small, they are prone to be maltreated, dropped and exposed to the most unfriendly of environments, which is probably the source of very many camera failures. The next most common source of problems is associated with electrolytic capacitors that dry up, become electrically leaky or leak electrolyte over the PCBs to produce corrosion and track breaks or even short circuits. The cleaning up process after such a disaster warrants the use of an ultrasonic cleaning bath which is a further expensive addition to the workshop equipment list.

The usual tests with a prerecorded tape can be used to isolate record from replay faults and the same mistracking problems of the VCR may be found in the camera. The use of plastic for many of the components associated with the cassette loading mechanism makes this area particularly prevalent to user abuse. Jammed cassettes are one area where the owner will often try to recover the unit first and then create even more damage. Although replacement parts may be relatively cheap, the extent of the labour in dismantling to effect the repair and reassembling afterwards can be costly.

The alternate use of a spray freezer and hot air gun can be used with advantage to localize those faults that seem to be temperature sensitive, i.e. those that appear after being in operation for a time. Interconnecting multiway cables can easily develop high-resistance joints that are often very difficult to detect. In particular, beware of soldered leads from ribbon cables that terminate on pins embedded in plastic – these usually melt very easily. Because of the possibility of rough handling, interconnecting multiway cables and sockets are often secured with a glue line that can with time become oxidized and conductive.

CHAPTER 10

TELEVISION DISTRIBUTION SYSTEMS OR COMMUNITY AREA TELEVISION (CATV)

Cable distribution systems have their origins in the late 1940s when it was found difficult to distribute good quality radio and television signals into cities and built-up areas. These systems initially distributed typically three radio and one television channels and much of the early work originated in the USA. The systems were originally built around multiway twisted pair cables using carrier frequencies as low as 5 MHz. As the systems expanded to provide more channels, it was necessary to change over to coaxial cable with carrier frequencies in the VHF range. Such networks became well established and eventually used carrier frequencies ranging up to about 450 MHz, finally up to the hyperband of 860 MHz (VHF and UHF bands). Currently the state of the art broadband networks now include television, radio, telephony and digital data services, including subscriber interactivity within their tier of services. These, of course, now employ optical fibre as the main backbone or trunk part of the network.

The major advantage of a cable system is its relative immunity to noise compared with over-the-air distribution. For any given bandwidth, the resolution of TV images is a significant improvement over normal aerial signals. Again for digital data services, the bit rate can be significantly higher for the same bit error rate (BER) relative to conventional broadcast services, a feature that is particularly valuable for interactive services. In spite of this advantage, it is important that the transmissions start out with a high signal-to-noise (S/N) ratio because the attenuation factor of the cable is also the level of the noise component added between amplifiers which themselves also add further noise to the system. Furthermore, impedance mismatch also introduces a noise component so that it is important that the matching along the network is maintained even when subscribers (subs) are added or dropped.

NETWORK TOPOLOGY

The ealiest systems were built up around the tree and branch network structure as indicated by Figure 10.1. The main *trunk* section of the network consists of typically 25 mm coax cable with a number of trunk amplifiers or repeaters spaced at intervals of rather less than 1 km. As the network approaches the subs premises, the coax cable progressively reduces to about 5–6mm diameter for the shorter final drops. Figure 10.2 shows some of the typical elements of a tree and branch network. Bridging amplifiers are used to feed to the branches from the main trunk maintaining the signal level and impedance matching. Both trunk and bridging amplifiers require a DC power supply and this may be fed directly from a suitable site, or over the cables if such is not available. The splitter element is used to extend the network and separate the trunk section to provide two equal signal levels. This component is usually a passive device that introduces some signal loss. At locations close to the subs, the tap element diverts the signal to each termination via the final drop cable. The main aim of the system design is to deliver TV signals with levels of about 1–3 mV amplitude

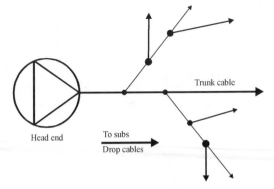

Figure 10.1 Tree and branch cable network.

for each channel and to each sub with the minimum of signal degradation.

Because the tree and branch system delivers all head-end signals to all subs, interactive and pay-per-view services are more difficult to provide. Since all communications have to be carried out through the head-end, the network quickly becomes bandwidth limited. Switched star networks were developed to overcome most of the problems associated with the tree and branch systems. The general principles of this configuration are indicated in Figure 10.3 where all subs connected to a *hub* head-end are also connected to each other via a network of electronic cross-point switches in the hub. In fact Figure 10.3 shows the multistar configuration that is commonly used for large area networks. Such a network can be fully interactive and provide radio, audio, television and data services that are uniquely encrypted for delivery to any authorized sub. Such services that are designed as broadband networks are usually based on a backbone of optical fibre.

THE HEAD-END

Figure 10.4 shows the general way in which all the various signals are organized before transmission to the network. The basic carrier

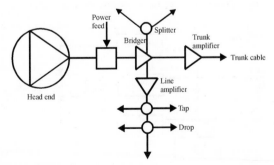

Figure 10.2 Elements of a tree and branch cable network.

217

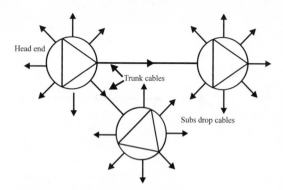

Figure 10.3 Switched star cable network.

frequencies are produced by a comb generator that creates carriers spaced by the normal terrestrial channels spacing –8 MHz in the case of the UK. The signals are modulated on to the appropriate carriers before being combined in the final transmission multiplex. In this way, locally generated signals may be added to the wider range of national channels.

THE TRANSMISSION MEDIA

25 mm diameter coax cable has an attenuation of around 3 dB per 100 m at 400 MHz, while for the subs drop cable these figures are nearer to 15 dB per 100 m at 400 MHz. As would be expected, the signal level to any particular sub then depends upon the distance from the head-end. For example, using the above data, a sub connected via a 200 m drop cable to a trunk of 4 km would experience a signal level loss of 30 + 120 = 150 dB. Thus a significant degree of amplification is necessary with a coax distribution system. The signal level along the network must therefore not be allowed to fall too close to the noise floor before a repeater is installed. Coaxial cable systems are also plagued with the ingress of moisture and therefore all cable inputs to the amplifier cabinets need to be adequately sealed.

Optical fibres consist of a core of high refractive index glass encased

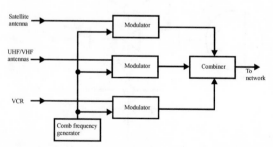

Figure 10.4 Head-end signal processing.

in a cladding of glass with a lower refractive index. As a protective measure, the cladding is further encased by a plastic coating. When light energy is launched into one end of the fibre, the wave-like motion of the light reflects between the core and cladding interfaces so that the energy propagates through the fibre. The driving source is usually produced by a solid state laser and the energy is detected by a light sensitive diode or transistor. As indicated by Figure 10.5, there are three basic types of fibre in use and each has specific characteristics that may be employed under different conditions.

In general, the outer diameter including the plastic coating is about 250 μm and the outer cladding diameter about 125 μm. The multi-mode step index fibre shown in Figure 10.5(a) has a core diameter of about 50 μm so that when light is launched into the fibre the energy can travel along a multiple of paths as indicated. This effect causes pulse dispersion which limits the shortest wavelength that can propagate. The losses, while small, are nevertheless highest of all the fibres. Thus multimode fibres are limited to relatively small bandwidths and short haul network systems.

By comparison, the monomode fibre with its core of about 10 μm diameter, shown at Figure 10.5(b), has only one mode of propagation so that the losses are very small, the bandwidth much wider, the transit time rather less and so is particularly suited for long haul, wideband networks.

The so-called graded index fibre with a refractive index which is highest at the centre and decreases across the radius of the core has a

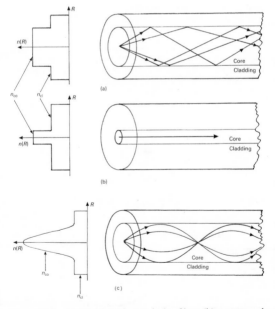

Figure 10.5 (a) Multi mode step index fibre; (b) mono mode step index fibre; (c) Graded index fibre.

typical diameter of about 50 μm. The rays therefore reflect off different radii and travel along a sinusoidal path, with the fastest waves taking the longer route. Thus all waves have virtually the same transit time giving the fibre the widest bandwidth, but at greater cost. Again, this type of fibre is suited to long haul, wide bandwidth networks. Fibre lengths for all three types can readily be jointed using a glass fusion splicer technique to give very low losses.

Optical fibre systems use one of two transmission windows, either between 1.3 and 1.35 μm or 1.5 and 1.6 μm. In either case, the signal level losses can be as low 0.2 dB/km, with jointing losses of less than 0.15 dB/joint.

In cable TV networks that employ optical fibres, it is common practice to counter the signal losses by periodically converting the light energy signal back into its electronic form for amplification before being relaunched back into the cable network. However, for very long haul networks that are encountered by the telecommunications industry, amplification is often carried out with the signal in the light energy state. Since what happens in one part of the communications business invariably spins off into the other branches, the concept of the erbium-doped amplifier is described here. Figure 10.6 shows the general principle of this technique. A loop of about 20 metres of silica glass fibre is doped with the tri-valent element erbium and linked into the network by splicing as indicated. The low-level signal to be amplified is combined with a high-level pump signal using a coupler. As the combined signal passes through the doped fibre, energy transfers from the pump through stimulated emission to amplify the wanted signal. Networks have been successfully constructed with such amplifiers spaced about 2000 km apart.

Optical fibres are commonly constructed in bundles of 8, 16 or 24 fibres around a metallic strength wire that is used for pulling the structure into network ducts. The metallic conductor can also be used to provide power for the line network. Such fibres are also affected by the ingress of moisture when the losses can increase quite dramatically due to the action of OH ions. These cables must therefore be adequately sealed. The bandwidth that is available can support either 40 HDTV channels, 200 conventional PAL channels or 30 000 telephone calls. Since each carrier occupies a different wavelength of light, the system is often referred to as *wavelength division multiplex* (WDM).

NETWORK PROBLEMS

The major problems associated with cable network systems are related to road-side cabinets that carry repeater amplifiers and distribution

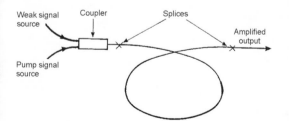

Figure 10.6 Erbium doped fibre amplifier.

points and the cables themselves. The former is a regular target for vehicle accidents and the latter is often dug up while contractors are repairing other buried services. However, the characteristics of both coaxial cable and optical fibres are so well known that changes of impedance along the network can readily be detected. In particular, both open and short circuits give a very large reflection of any energy launched into the cable. Knowing the propagation constants it is a simple matter to determine the time lapse between transmitting a pulse to the faulty network and receiving its reflection back at the source. Using this time domain reflectometry (TDR) technique which can be applied to either metallic conductors or optical fibres, the distance to the fault can readily be calculated.

MASTER ANTENNA TELEVISION (MATV)

Smaller networks designed to service buildings and apartment blocks are referred to as MATV systems and are invariably based on coax distribution. The channel's inputs are obtained from aerials located on the roof and distributed over parallel paths via a distribution amplifier mounted near to the aerial. The outputs of several aerials can be combined to provide VHF-FM radio and satellite and terrestrial television services. The distribution network includes diplexers and splitters, some of which may be active devices.

For large, single family houses that need several TV sets, VCRs and Hi-Fi audio distribution, a similar small version of the MATV system can be usefully employed. However, unless care is taken in the original design and development of these systems, the adding or dropping of receivers can create some anomalies with impedance matching. The outputs from two or three aerials can be combined with diplexers or triplexers and their frequency selective filters can be valuable in preventing cross-modulation between feeds. When designing any remodulation system it is important to establish a coherent set of channel spacing to avoid cross-modulation. Many of these problems are well documented in *Television* articles. Satellite master antenna television (SMATV) systems are special branch MATV systems and are more difficult to design, simply because of the higher carrier frequencies involved and the need to use high-grade low-loss feeder from the LNB head-end. In fact these systems seem to be more highly susceptible to the ingress of moisture into the feeder cable. Such extended systems often cause problems of impedance mismatch when new services or receivers are added or dropped at a later date. With MATV systems it is possible to find ghosting caused by pre- or post-echoes when a strong local signal breaks through into one or more receivers tuned to a specific channel. The usual cure in this case involves either moving the master aerial or retuning the system to frequencies that avoid the problem.

Apart from maintaining a high S/N ratio for MATV systems, it is important to be aware of the effects of the following possible signal degradations: luma/chroma gain inequality and cross-talk and differential phase and gain.

ASYMMETRICAL DIGITAL SUBSCRIBERS' LINE (ADSL) NETWORK

Throughout the world there are many millions of telephone twisted pair cables extending from exchange buildings to subscribers' premises.

These represent the installations that provide normal voice traffic using copper or aluminium wires in various gauges, ranging from 0.32 mm to 0.63 mm in diameter. Any particular subscriber's link may consist of more than two of these in series so that the line attenuation may vary from link to link.

It has been shown experimentally that these installations, originally designed for low voice frequencies (300 to 3400 Hz), are capable of carrying much higher frequencies, provided that the line characteristics are suitably countered. Figure 10.7 shows approximately how the line attenuation varies up to 1 MHz. Over this same range the group delay which represents the ratio of phase change per unit Hz is almost constant. This means that because the phase change introduced by the line is proportional to the signal frequency, there will be little pulse distortion and consequently fewer bit errors in a digital transmission.

These line characteristics simultaneously allow a multiplex of signals including high bit rate digital transmissions over voice grade cables. Using MPEG digital compression, video signals of various grades can be transmitted direct to the home. A bit rate of 2.048 Mbit/s can provide VCR quality video together with digitally coded stereo sound. At 6.144 Mbit/s the video quality is equal to that of a broadcast transmission. The bandwidth provides for a low bit rate user return channel for control purposes, ranging from 9.6 or 16 kbit/s up to 64 kbit/s. The 6.144 Mbit/s downstream channel can be allocated in various ways. 4 × 1.536 Mbit/s, 3 × 2.048 Mbit/s and 1 × 6.144 Mbit/s, with other combinations possible. However, the higher bit rate systems tend to have a shorter network penetration. ADSL can provide video with VCR type characteristics such as fast forward, fast rewind, freeze frame, and pause, etc., on demand. The general channel allocation is indicated by Figure 10.8 and as a consequence of the availability of the telephone system, there are many trials in operation around the world. Adaptive separation filters are employed at each end of an ADSL network in order to combine or separate the services with the minimum of mutual interference. Digital adaptive equalizers adjust automatically to suit individual line characteristics which include temperature- and moisture-dependent variable parameters and continuous interference. Provision is made so that at each new acquisition, the characteristics of the line amplifiers and equalizers are reset under a short training sequence to establish the maximum signal-to-noise ratio.

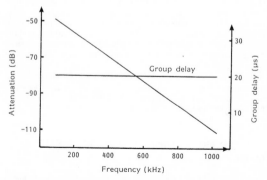

Figure 10.7 Telephone line characteristics.

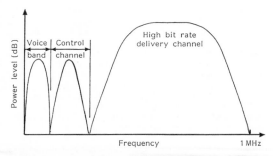

Figure 10.8 Channel allocation for ADSL.

In order to maximize the channel capacity a form of bit rate reduction coding known as 2B1Q is employed. With this, each pair of binary input digits is converted into one of four quaternary symbols. Further, to minimize the effects of both radiated and induced radio frequency interference, ADSL employs a balanced feed of RF signals to the line.

Three different modulation techniques are under test in the various trials.

Quadrature amplitude modulation (QAM)

The binary data stream is first split into two substreams and each separately modulated on to orthogonal (in this case quadrature) versions of the same carrier. These two modulated signals are then added before transmission to line.

Carrierless amplitude/phase modulation (CAP)

The bit stream is again split into two components as shown in Figure 10.9 and then separately passed through two digital filters that have an impulse response differing by 90°. The outputs are then added, passed through a digital-to-analogue converter and filtered before being passed to the transmission network.

Discrete multi-tone modulation (DMT)

This preferred method has a lot in common with COFDM (coded orthogonal frequency division modulation), in that the main channel is subdivided into many sub-channels. Each serial digital input signal is first encoded into parallel format and then passed through a fast Fourier transform processor to convert the frequency domain samples

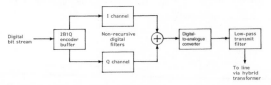

Figure 10.9 Bit stream coding for CAP modulation.

into time domain values with a sliding time-window effect. The 2-bit prefix is included to break up the bit stream for the receiver and therefore more accurately define the Fourier window. The values are then transcoded into a serial format, then digital-to-analogue converted before transmission. The general principles of this form of processing are shown in Figure 10.10. The line characteristics for each narrow subchannel are practically constant so that the minimum of pulse smearing is created thus improving signal quality. Any impulsive noise that is input as interference is spread over many subchannels by the FFT processor window so that this form of interference is less likely to create data errors. The number of bits transmitted in a subchannel can be varied adaptively depending on the signal-to-noise ratio in any channel. This not only improves the signal quality on a particular line, but it also minimizes the effect of cross-talk from other lines. The specifications for the system includes error control in the form of Reed-Solomon coding with the possibility of including interleaving.

Hardware development is aimed at a single board approach with extensive use of standard VSLI chips wherever possible and with consumer prices very much in mind. Data rates in excess of 6.144 Mbit/s have already been tested over relatively short voice grade transmission lines and the system is readily adaptable to use over optical fibre networks. The technology is the subject of both the American National Standards Institute (ANSI) and the European Telecommunication Standards Institute (ETSI).

A WIRE-LESS DISTRIBUTION SYSTEM

There are many valley areas where it is impossible to receive either television pictures or stereo radio satisfactorily. Often these represent small communities which make it uneconomic for these services to be provided via the national system or a cable network. In addition, there are many areas of towns and cities where it is not economical or even feasible to bury some sections of a cable network. A system known variously as Multichannel Multipoint Distribution Services or Multichannel Microwave Distribution Services (MMDS) has for some time been used in the USA and Canada to provide broadband services. MMDS can then provide an effective solution to these problems. Further acronyms for these services include MVDS (Microwave Video Distribution Service) and M³VDS (Millimetre-wave Multichannel Multipoint Video Distribution System).

In its original form, MMDS operated in the 2.5 GHz band, to provide an educational service to rural areas or to retransmit either national or satellite-sourced TV signals. A similar service has been successfully established to provide an extension to the 12 channel cable network in County Cork, Eire, and this came into service during 1990. The frequency range from 2.5 to 2.686 GHz provides for 23 PAL I system channels. These are operated using carrier frequency off-set plus vertical or horizontal polarizations to minimize interference. The

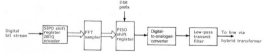

Figure 10.10 Bit stream coding for DMT modulation.

23 channels are utilized in such a way as to provide service from at least 11 TV transmissions for all cable subscribers. Because of the possibility of interference with other services at 2.5 GHz, the UK broadcasting authorities have carried out trials using the 29/30 GHz band. MMDS is essentially a local wideband rebroadcast system for the delivery of Video, including teletext; sound, including stereo; and data, over a restricted area. High-power MMDS using transmitter output powers of 10 to 100 watts can provide a service extending to more than 40 km. Low-power MMDS (1–10 watts output) can provide coverage for up to 10 km, depending on the power output, transmit antenna height, and the terrain, as a line-of-sight path is necessary. Low-power MMDS tests have been carried out and reports show that for a transmitter RF output power of 1 watt, TV pictures of excellent quality can be obtained at distances in excess of 5 km.

Figure 10.11 shows the basic principles of the operation of an MMDS system. At the transmitter, the incoming signals are either up- or down-converted as required, filtered and then amplified at 2.5 GHz. Earlier high-powered transmitters used one driver/power amplifier stage per channel. However, recent GaAs FET developments now allow the low-power systems to combine channels into blocks for delivery via a single driver/power amplifier stage. The provision of power supplies for rebroadcast transmitters at isolated sites is always a problem. With low-power MMDS it is possible to use the latest technology of battery operation, with recharging being achieved from either a wind or solar generator. At the receiver, the LNB down-converts the 2.5 GHz band signals to frequencies that allow direct input to standard receivers.

The antennas used at 2.5 GHz operation for both transmit and receive are often around 50 cm diameter, or of similar rectangular area. Because of the relatively low microwave frequencies involved, the reflectors can be perforated to reduce windage and lightly corrugated to increase the mechanical stiffness, without introducing significant surface error losses. The reflectors can easily be fabricated from stainless steel for good corrosion resistance. Such antennas and their mountings are less expensive than Yagi arrays of a similar gain. A 50 cm diameter dish typically has a beamwidth of 15° so that pointing

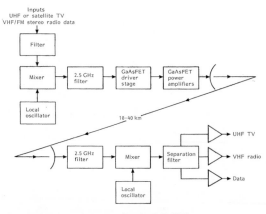

Figure 10.11 Basic processing for MMDS systems.

225

accuracy is no problem; neither is antenna movement with wind. A wedge of more than 2.5 km wide can be covered with a beamwidth of 15° at a range of 10 km. This represents a very significant spread of signal energy. By comparison, the antennas used for 29/30 GHz operation can be as small as 300 mm diameter.

There are many ways in which the retransmission bandwidth can be used. The simplest being a block translation of the incoming channels. However, in some cases, this may be wasteful of spectrum. Sometimes the four UK UHF TV channels are spread over 128 MHz, although the spread is more usually 80 MHz, but Channel 5 could produce further problems. The retransmission of VHF FM stereo, including local radio, would require a further 14 MHz for five UK channels. Thus demodulation and remultiplexing would appear to be one answer to this problem. This would then provide enough spectrum for some satellite-delivered TV as well as data services.

Simple link power budget calculations show the viability of including a digital data service within the frequency multiplex. It is thus possible to provide end users with a digital return channel, giving them a full duplex data service. Since only about 10 mW of RF power is required for a low-speed digital return path, this could easily be provided by a fairly simple Gunn diode transmitter circuit.

INDEX